FOOLED

by the

WINNERS

FOOLED

by the

WINNERS

How Survivor Bias Deceives Us

DAVID LOCKWOOD

GREENLEAF
BOOK GROUP PRESS

This book is intended as a reference volume only. It is sold with the understanding that the publisher and author are not engaged in rendering any professional services. The information given here is designed to help you make informed decisions.

Published by Greenleaf Book Group Press
Austin, Texas
www.gbgpress.com

Distributed by Greenleaf Book Group

For ordering information or special discounts for bulk purchases, please contact Greenleaf Book Group at PO Box 91869, Austin, TX 78709, 512.891.6100.

Design and composition by Greenleaf Book Group
Cover design by Greenleaf Book Group
Cover Image: ©iStockphoto/Andrew_Howe

Publisher's Cataloging-in-Publication data is available.

Print ISBN: 978-1-62634-880-6

eBook ISBN: 978-1-62634-881-3

Part of the Tree Neutral® program, which offsets the number of trees consumed in the production and printing of this book by taking proactive steps, such as planting trees in direct proportion to the number of trees used: www.treeneutral.com

TreeNeutral

Printed in the United States of America on acid-free paper

21 22 23 24 25 26 27   10 9 8 7 6 5 4 3 2 1

First Edition

To my family

"The armor (on the returning planes) doesn't go where the bullet holes are. It goes where the bullet holes aren't."

—JORDAN ELLENBERG,
referring to Abraham Wald's conclusions

"The cemetery of failed restaurants is very silent."

—NASSIM TALEB

CONTENTS

Ships, Sailors, and Prayers

Diagoras and the Existence of the Gods

The first person in history known to have written about survivor bias was the 5th-century BC Greek philosopher Diagoras of Melos.[1] Diagoras was a traveling poet who earned commissions from wealthy patrons by writing hymns to the gods.

Diagoras is most famous as the "first atheist." Ancient manuscripts reveal that Diagoras once wrote a poem that was stolen by another poet, who went on to achieve great success with the composition. Diagoras confronted the thief, who swore a solemn oath to the gods that the work was his. Furious at the man's refusal to admit to plagiarism, Diagoras implored the gods to punish the man for the appropriation of another's creative efforts and for swearing a false oath. When this demand for retribution went unanswered, Diagoras lost faith in the Greek gods.

Diagoras was challenged by a friend to defend his atheism. The friend pointed to paintings on the walls of the temples of sailors who were saved from violent storms after praying to the gods.

The friend then asked: "You think the gods have no care for man? Why, you can see from all these votive pictures here how many people have escaped the fury of storms at sea by praying to the gods who have brought them safe to harbor."

Diagoras replied: "Yes, indeed, but where are the pictures of all those who suffered shipwreck and perished in the waves?"[2]

As a result of this exchange, Diagoras secured his place in history as the first atheist and the first to explicitly call out survivor bias. In this example, Diagoras's friend focuses on the survivors of the storm. Those who perished despite their pleas to the gods were no longer among the living, and thus their images were not portrayed on murals throughout Athens.

Diagoras's friend was fooled by the winners. By focusing only on those who survived the storms at sea and ignoring those who drowned, his friend was led to the wrong conclusion about the existence of the gods.

Survivor Bias

We are fooled by the winners because survivor bias distorts our thinking. Survivor bias has been defined as "a logical error of concentrating on the people or things that made it past some selection process and overlooking those that did not."[3] In other words, we fail to learn from the non-survivors, those who have lost.

This tendency to focus on the winners is understandable. We are fooled by the winners because the winners, whether people, objects, data, or ships that return safely to port, are often easier to observe. But in many instances, we can derive as much or even more insight from those who have lost.

I began to write this book after spending three decades working on Wall Street and in Silicon Valley, witnessing firsthand how survivor bias was used to deceive us. As I continued to investigate, I discovered that the deceptive influence of survivor bias on our thinking extends beyond what I had first imagined.

Through the process of researching and writing this book, I uncovered a number of unexpected instances in which survivor bias impacts our daily lives. I also became convinced that survivor bias clouds our view of the past, which prevents us from gaining a clear view of our future.

Two Types of Survivor Bias

Survivor bias comes in two types that are differentiated by our ability to observe the non-survivors.

The first could be called "Diagoras"-type survivor bias. In the example of the Greek ships, Diagoras is the observer. He is able to count the number of ships that returned safely and those that did not. Even if all the ships floundered and were lost at sea, Diagoras could still observe that all the sailors had drowned.

The second could be labeled "Sailor"-type survivor bias. In this case, the observer is a sailor on one of the ships. A sailor is able to count the number of ships that left port and safely returned, but only if he survives the voyage. Just as in the case of Diagoras-type survivor bias, a sailor can be deceived upon returning to port by failing to consider the non-surviving ships. But he can be further misled if he does not realize that his observations are limited to only those instances in which he is still alive.[4]

About This Book

This book is split into two parts. In Part I, we will look at Diagoras-type survivor bias. We will see how survivor bias is used by others to deceive us into overpaying hedge fund managers, taking pills that don't cure our ills, following the bad advice of many advice books, overspending on lotteries, opening restaurants that are likely to close, and suffering through diets that don't help us lose weight. We will show how a professor discovered during WWII that Allied Bomber Command was deceived by survivor bias, and that even today history is still written by the winners.

In Part II, we will examine Sailor-type survivor bias. We will demonstrate how survivor bias distorts our perspective on evolutionary history, including our good fortune that 65 million years ago a space rock targeted our planet, wiping out the dinosaurs, and that against all odds fifty thousand years ago we vanquished at least five other human species. We will analyze how survivor bias misleads us about the dangers of atomic weapons and global warming. We will also see how survivor bias aids the resolution of Fermi's paradox (the puzzle concerning the existence of intelligent life in the universe) and provides insights into the mortality of Schrödinger's cat (one of the most important questions in modern physics).

Previous studies of survivor bias have been inaccessible to most, housed in formula-laden statistical journals or tucked away in academic papers written for subject area experts. But you won't find any math or technical jargon here. Instead, we will apply the concept of survivor bias to concrete, real-world examples. Just because math is my favorite subject does not mean it should be yours. We will put flesh on the bones of this theoretical concept by working through examples—minus the equations. As each example is discussed in turn, we will examine how survivor bias is used to deceive us and how we manage to deceive ourselves. Then, we will set out ways to correct for survivor bias so we are not fooled by the winners.

We begin with survivor bias in financial services. The first person to call out how survivor bias is used to deceive us in financial services was a New York stockbroker who wrote a book in 1940 that remains to this day one of the most quoted texts on the subject of Wall Street.

The stockbroker began his book with a question about boats.

PART I

CHAPTER 1

Financial Services:
So Much for So Little

Fred Schwed and Customer Yachts

Fred Schwed (1902–1966) was, in the terminology of his day, "a customer's man." For almost two decades, he worked for the Wall Street firm of Edwin Wolff, hawking stocks to retail investors. Today, he would be called a registered representative, or more commonly a stockbroker. In his spare time, he wrote books, including a popular children's novel, *Wacky, the Small Boy*, a bestseller during the 1930s.

Schwed was the son of a member of the New York Curb Exchange, later renamed the New York Stock Exchange. In 1924, after graduating from Columbia University, he went to work on Wall Street. The 1920s was a period of rampant speculation, when the allure of a booming market and cheap margin debt seduced millions of individuals to place their life's savings into stocks. After the equity markets crashed in 1929, Wall Street and the nation were plunged into a depression that lasted until WWII. In fact, the Dow Jones Industrial Average (DJIA) did not return to its 1929 height in absolute terms until 1958. So, during the many years he worked on Wall Street, Schwed experienced the best and worst of times.

Schwed developed a keen eye for the foibles and follies of his colleagues and shaped his observations into pithy anecdotes. In 1940, he cast his thoughts into a humorous and insightful short book entitled *Where Are the Customers' Yachts? or A Good Hard Look at Wall Street*.[1] The title of Schwed's book comes from his observation that Wall Street is a more profitable place to work than to invest.

Schwed wrote:

> *Once in the dear dead days beyond recall, an out-of-town visitor was being shown the wonders of the New York financial district. When the party arrived at the Battery, one of his guides indicated some handsome ships riding at anchor. He said,*
>
> *"Look, those are the bankers' and brokers' yachts."*
>
> *"Where are the customers' yachts?" asked the naïve visitor.*[2]

Today, the financial services industry continues to be a lucrative place to earn a living. In Schwed's words, "Wall Street is the highest paying spot on the face of the Earth."[3] The financial sector of the US economy employs fewer than 5 percent of all US workers yet claims almost 30 percent of all corporate profits.[4]

The financial sector is extraordinarily profitable due to the economics of its primary business: managing the savings and investments of the nation, or what is known as asset management. For US securities firms, asset management is critically important, accounting for just over half of total revenue.[5]

The asset management business is extraordinarily profitable partly because of the way it charges investors. Unlike many other service industries, most financial services professionals are paid on a percentage basis. By contrast, lawyers bill per hour, regardless of the size of the transaction. This makes sense: The cost of a lawyer's time is the same whether the deal to be negotiated is for $100 million or $1 million. But charging a percentage of assets under management is the most common fee arrangement for fund managers. For instance, hedge funds typically collect fees of 1 to 2 percent of assets under management and 20 percent of all investment gains. Hence, a hedge fund manager charges one hundred times more for investing $100 million than $1 million. The same applies to wealth advisors and mutual fund managers, who generally charge a fee based on a percentage of assets.

The exceptional profitability of the asset management business is reflected in employee compensation. The average mutual fund portfolio manager earns about $1.3 million in salary and bonuses.[6] At a hedge fund, that number is $1.4 million.[7] Those at the very top of the US hedge fund industry are paid even more. In 2017, the twenty-five highest compensated hedge fund managers were paid an average of $615 million each.[8] Assuming a 250-day work year and ten-hour days, that equates to $246,000 per hour. In the United States, the average worker earns an hourly wage of $27.[9] These unusual levels of compensation can make headlines: A prominent hedge fund manager was reported to have purchased a second home in New York for $250 million. No similar purchases are known to have been made by his investors.

The justification for the extraordinary amounts of money paid to hedge fund managers is that they outperform the market indexes. There would be no reason to pay a hedge fund manager to match market returns; an index fund achieves that result without the exorbitant fees. To convince investors to pay these fees, hedge fund managers offer "proof" of their ability to generate returns in excess of market benchmarks. But the proof put forward is riddled with survivor bias. Schwed warned us of this in 1940.

He wrote:

> *Figures, as used in financial argument, seem to have the bad habit of expressing a small part of the truth forcibly, and neglecting the other part, as do some people we know.*[10]

John Meriwether and LTCM

John Meriwether is an example of how the failure to adjust for survivor bias benefits hedge fund managers.[11]

Meriweather grew up in Chicago and from an early age was an avid gambler. As a teenager, he bet on horses, blackjack, baseball, and golf, a sport in which he excelled. He graduated from Northwestern University, taught high school math for a year, and then went to the University of Chicago, earning an MBA in 1973. He was hired by Salomon Brothers after business school and joined the fledging fixed income department. In 1977, he founded the arbitrage group within Salomon Brothers. The name was apt. Rather than betting on the absolute price change in a security, Meriwether specialized in betting on the change in the relative price difference between two fixed income instruments, or what is known as bond arbitrage.

In the late 1970s and 1980s the arbitrage group at Salomon Brothers enjoyed a number of successful years and generated a substantial portion of the firm's profits. Meriwether soon climbed to one of the top rungs of the Salomon Brothers' corporate ladder.

However, in 1991, Meriwether and his arbitrage group submitted a series of false bids to the US Treasury in order to gain a disproportionate share of newly issued securities. The resulting crisis almost bankrupted the brokerage firm.

To restore faith in the firm, Warren Buffett stepped in as CEO and Meriwether resigned, although Meriwether defended his actions and those of the arbitrage group to the very end. The SEC charged Meriwether with the civil crime of failing to supervise his employees, and he settled with the SEC, agreeing to a three-month suspension from the securities business and a $10,000 fine. After he left, Salomon Brothers recovered, and the firm was sold to Citigroup in 1997.

After departing Salomon, Meriwether founded Long Term Capital Management (LTCM). His pitch to investors was that he would undertake the same arbitrage strategies at LTCM that had proven successful at Salomon Brothers. He also recruited a number of leading academics and brought many of his former colleagues from Salomon Brothers to the new hedge fund. Some were surprised that Meriwether, after being fined and suspended from the securities business by the SEC and almost bankrupting his firm, could raise monies from investors. However, he persuaded enough individual and institutional investors to trust him with their savings to open his new fund in 1994 with over $1 billion in capital. (He pitched Warren Buffett to invest, but Buffett declined.)

From 1994 to 1997, the returns on investor capital at LTCM substantially exceeded the market indexes, and the firm steadily grew in size. By the early months of 1998, total capital under management had risen to $4.7 billion.

And then it all went pear-shaped.

During 1998, the Asian financial crisis weighed heavily on the debt and other financial markets. Returns for LTCM turned sharply negative in May of that year and continued to be negative throughout the summer months. In August, Russia declared a debt moratorium, effectively suspending payment on Russian-issued government bonds, sending emerging market debt into a tailspin. Investors around the world fled to the safety of more liquid, higher rated securities. By the end of August, LTCM had lost almost $2 billion in capital.

LTCM lost money primarily because the firm had been betting that the spread would narrow between higher rated, more liquid securities and those with a lower rating and less liquidity. From the founding of LTCM in 1994 to 1997, LTCM's bet paid off as the difference in price between these two groups of securities had steadily declined. However, during times of crisis, investors flee to the security of higher rated and more liquid securities. The summer of 1998 was no exception, and the spread between these two groups of securities significantly widened. By September, LTCM had lost another $2 billion and was insolvent.

In response, the Federal Reserve Bank of New York took control of LTCM from Meriwether and liquidated most of his investments. LTCM investors were almost entirely wiped out, and the banks that had lent money to the firm incurred substantial losses. LTCM's largest creditor, UBS, lost $780 million, and its chairman was forced to resign.

Not surprisingly, LTCM did not report results to the hedge fund indexes for 1998. If LTCM had reported results, investor returns would have been a negative 91 percent for that year.[12] During the period in which the firm was part of the hedge fund indexes, from 1994 to 1997, LTCM reported that investors gained an average of 32 percent per year.[13] However, from 1994 to 1998, including returns from that last nonreporting year, LTCM lost investors an average of 27 percent per year.[14] From 1994 to 1997, the returns of LTCM were included in the major hedge fund indexes. Once LTCM closed up shop, performance of the fund was deleted from the historical returns of those indexes.

Survivor Bias and Hedge Fund Performance

LTCM is an example of how investors evaluating the performance of hedge funds as an asset class can draw the wrong conclusions due to survivor bias. The returns reported by a hedge fund index contain only the past performance of the surviving funds. The non-survivors, such as LTCM, are excluded from the indexes. A hedge fund index should really be called an "Index of Not-Failed Hedge Funds."

Hence, due to survivor bias, hedge fund indexes overstate the performance of hedge funds as an asset class. This is particularly true because the rate of hedge fund failure is high. If only a small percentage of funds ceased operations each year, the effect of survivor bias on hedge fund indexes would be minimal. But this has not been the case.

In a study of 720 hedge funds that started twenty years ago, only 262 were still operating after the first decade, and only 13 after the second.[15] Another study of over one thousand hedge funds showed that from 2004 to 2014, fewer than half that started at the beginning of the period were still in business ten years later.[16] As expected, the number of hedge fund closures varies with market conditions. One survey showed 16 percent of hedge funds had dissolved in 2007, but that number increased to 31 percent in 2008 during the financial crisis.[17]

And not all hedge fund closures are involuntary. The compensation structure of hedge funds provides an incentive for a hedge fund manager to close up shop after a bad year.

Hedge Fund Compensation: Never Having to Say You're Sorry

Investors typically agree to pay hedge fund managers 20 percent of investment gains each year, or what is known as carried interest. (In financial terms, a hedge fund manager receives a free call option that can be exercised each year, based on performance over the previous twelve months.) But if a fund manager loses money one year, then those losses have to be made up before additional carried interest can be paid. This feature of hedge fund compensation is known as a high-water mark.

An example illustrates this point.

Suppose a hedge fund manager is given $100 to invest. In Year 1, the portfolio gains 20 percent and is now worth $120, a gain of $20. The hedge fund manager will typically receive 20 percent of the profits, or $4 of carried interest in compensation for Year 1. In Year 2, the portfolio declines by 50 percent and is now worth $60. There are no profits, so the hedge fund manager receives no carried interest.

But the high-water mark remains at $120. Even if in Year 3 the value of the portfolio doubles to $120, the manager will receive no carried interest. Additional carried interest will be paid only if the portfolio becomes worth more than $120 in subsequent years. However, things still worked out just fine for the hedge fund manager since she gets paid at the end of each year. In our example, even though over two years she lost $40 of her investors' $100, she was still paid $4 in carried interest to do so.

Unfortunately for investors, after her disastrous performance in Year 2, there is an incentive for the hedge fund manager to bank her $4 and close up shop. From the perspective of the hedge fund manager, she will be working for free until the value of the investors' portfolio reaches the high-water mark of $120, an increase of 100 percent in the value of the portfolio. She is financially better off leaving to start a new hedge fund, unburdened with overcoming the hurdle of the old high-water mark due to her failed performance.

In the case of LTCM, Meriwether was forced out. But even if he hadn't been, he had little incentive to continue. With a loss of 91 percent during the final year, Meriwether would have had to work many years before he saw another bonus from carried interest, if ever. In fact, like many in the hedge fund industry who fail, Meriwether promptly started a new fund. In 2000, Meriwether launched JWM Partners with some of his former colleagues from LTCM. The fund was successful until 2008 but then suffered large losses during the financial crisis. JWM Partners was dissolved in 2009.

In 2010, Meriwether opened a new fund, JM Management.

Admittedly, not all hedge fund managers have been able to fail and then open new funds repeatedly like Meriwether. It is also difficult to determine what percentage of hedge fund closures are voluntary versus involuntary. In many cases the decision is probably mutual.

Regardless, as we have seen, the vast majority of hedge funds fail. This high rate of failure has led to significant survivor bias in the hedge fund indexes. For the ten-year period ending in 2014, survivor bias increased the nominal returns of one hedge fund index by 26 percent.[18] Another study exposed that from 1996 to 2003, survivor bias raised the performance of a different hedge fund index by 47 percent.[19]

Survivor Bias and Mutual Fund Performance

The indexes measuring the performance of mutual fund managers are similarly distorted by survivor bias.

Mutual funds are subject to the same high rates of failure as hedge funds. During the period from 1997 to 2011, only 54 percent of mutual funds managed to survive.[20] Another study found failure rates of 78 percent to 95 percent for different types of mutual funds over a twenty-year period.[21] Not surprisingly, funds that go out of business underperform those that don't. One study of over five thousand mutual funds found that the surviving funds outperformed the non-survivors by more than 20 percent on average.[22]

As a result, mutual fund indexes overstate performance by excluding the returns of the non-survivors.

This is confirmed by comparisons of mutual fund performance and the market indexes with and without the non-survivors. A study of large-cap value funds found that 62 percent had beaten the market averages over the past five years.[23] However, once mutual funds that had gone out of business were included, the percentage of outperformers fell to 46 percent, no better than chance. The same study established that during the last fifteen years the percentage of outperformers fell from 50 to 27 percent with the inclusion of non-surviving funds. Another study examined 39,000 mutual funds over an even longer period—fifty years—and found that 13,000 had died.[24]

Individual investors are sold mutual funds on the basis of past performance. This past performance is typically calculated by your neighborhood stockbroker who, consciously or not, fails to correct for survivor bias in his sales pitch.

Stockbrokers charge fees to investors in exchange for a promise of better investment returns. A common means of collecting fees is to sell you a mutual fund run by a fund manager who is supposed to be better at picking stocks than you are. The fund manager ought to be better than you or the index; otherwise, there is no reason to pay two sets of high fees—one to the fund manager to pick stocks and another to your stockbroker for introducing you to a fund manager to pick stocks for you. After all, anyone could buy stocks through an online brokerage firm or invest in an index fund for a nominal charge.

To justify two sets of high fees, your stockbroker presents color-coded, bar-graphed, outside-audited financial statements demonstrating that over the past ten years the recommended fund manager has outperformed the market averages by wide margins.

Your neighborhood stockbroker claims this is due to the fund manager's superior stock-picking acumen.

If the stockbroker's firm started one mutual fund ten years ago, and that fund outperformed the market for ten years, then this particular fund manager may be highly skilled. However, if the stockbroker's firm started ten mutual funds ten years ago, and the other nine subsequently went out of business due to lousy performance, then you should include all ten funds in your assessment of the brokerage house's investment chops. It could be that the one remaining mutual fund manager was just plain good—or just plain lucky. It could also be true that the brokerage firm is not so good at picking fund managers, given that nine out of ten fund managers at the firm failed. Admittedly, the causes of the poor performance may be difficult to determine. But if the fund offered to you is the only one of ten funds still around, then the proposition that the manager of that fund possesses superior stock-picking skills is an example of the deceptive influence of survivor bias on our thinking. In many cases, what your neighborhood stockbroker is really selling is a mutual fund manager who has benefited more from good fortune than adept portfolio management. Our neighborhood stockbroker is touting one of the few surviving funds while passing over the more numerous non-surviving funds in silence. Your stockbroker hopes you will be fooled by the winners.

One way to test for survivor bias in mutual funds is to ask your stockbroker for a list of all the funds that he recommended ten years ago. Or ask for a report on the performance of the firm's mutual funds that includes those that were shut down over the last decade, the non-survivors. (I have done this myself. To date, no such reports have arrived in my inbox.)

Survivor bias also puts a spotlight on our natural tendency to attribute success to skill rather than luck. We want to believe that winning is due to ability and effort. In fact, success can be simply due to the law of large numbers. There are more than ten thousand mutual funds in the United States today, and some of them by pure chance will outpace the market indexes.

To illustrate, suppose all ten thousand mutual funds simply flipped a coin at the start of each year to determine whether to be long or short the market. After five years of coin flipping, we expect 312 of these funds to have consistently beaten the market every single year.[25] A much larger number will have outperformed the market over that period due to the gains in some years offsetting the losses in others. Of course, selecting one of the 312 funds that beat the market every year for the past five years would not be a formula for investment success. These funds are just the lucky survivors of a random process. Given that these funds owe their success to chance, the likelihood that these 312 funds will be successful in the future is no greater than that of other funds. In this example, future returns will not be correlated with past success.

This is exactly what studies of mutual funds have determined. One study reviewed actively managed mutual funds and found that "out of 687 funds that were in the top quartile as of March 2012, only 3.78 percent managed to stay there by the end of March 2014."[26] Another study looking at the period from 1996 to 2014 concluded that among mutual funds "there is no meaningful relationship between past and future fund performance."[27] The lack of statistical correlation between past and future performance confirms that outperforming funds largely owe their success to survivor bias. Somebody is going to win. It just happened to be them.

Correcting for Survivor Bias in Hedge and Mutual Funds

As we have seen, those in the financial services industry use survivor bias to deceive us. This cognitive failure is one way Wall Street, year after year, can sell products and services that promise more than they deliver.

The method to correct for survivor bias is to include the non-surviving managers and funds in the indexes in order to accurately measure performance of these asset classes. Furthermore, an asset manager's cumulative track record should be reviewed. Those who repeatedly start new firms should be judged by their lifetime performance.

This omission of the non-survivors is rarely seen in other financial market indexes. For example, the Dow Jones Industrial Average (DJIA) includes the performance of those companies that have suffered large declines in market capitalization or have been kicked out of the index. None of the original members of the DJIA are still part of the index today: General Electric (GE) in 2018 was the last of the original members of the index to be deleted. Nevertheless, the historical returns of the DJIA include the returns of those companies in the index, like GE, right up until the day of deletion. Not surprisingly, there is a correlation between stock market performance and getting booted from the index: The non-survivors typically underperform the index in the years before deletion. GE stock dropped 43 percent in 2017 and then another 56 percent in 2018. But the poor performance of GE while it was part of the DJIA is included in historical returns, unlike the hedge and mutual fund indexes.[28]

There is a simple fix for the hedge and mutual fund indexes: Calculate them in the same manner as other financial market indexes, such as the DJIA, by including the returns of the non-surviving funds. While this will not be greeted with enthusiasm by hedge and mutual fund managers, it will reflect the true performance of these asset classes. The same can be said for the indexes of other alternative asset classes, such as private equity, commodity funds, and funds of funds, which also do not include the non-survivors.

Fund Managers: Necessary but Overpaid

We have seen how those who work on Wall Street are highly compensated and use survivor bias to deceive us. Nevertheless, these individuals do play an important role in our economy. Professionals in the financial industry set the price of capital in order to funnel the savings of a nation into the most productive investments. The efficient allocation of capital is one of the main strengths of a market-based economic system. The price of capital determines what gets produced and what doesn't. If the price of capital is not set correctly, then goods and services can be produced that nobody wants. See Stalin's Soviet Union and Mao's China.

Without fund managers, there would be no one to perform this role. Placing all of a nation's savings into passive index funds would leave no one to price stocks. This is why a vibrant financial services industry is critical to economic growth, and the professionals in this industry should receive a fair wage. Of course, financial markets can overshoot during periods of rampant speculation or collapse in the aftermath of a market bubble. Even the most steely-eyed, savvy market professionals are not impervious to bouts of greed and fear. Those on Wall Street are flesh and blood human beings, despite reports to the contrary. Regardless of these shortcomings, a market-based system for capital allocation has been shown time after time to be better than all the others. I am not arguing that active asset managers are not necessary. I just think they are overpaid.

Fred Schwed was justifiably skeptical of the ability of those working on Wall Street to predict stock prices. Based on his experiences, he cautioned investors that:

> *The more skillful of these traders can and do peer into the future for a period of five or even twenty minutes concerning those securities which they are*

actively trading . . . However, as soon as they get a moment at leisure, they try peering into the future for five or ten months. For this they are precisely as ill equipped as everybody else.[29]

In concluding his book, Schwed wrote the following wry commentary:

Perhaps what you are looking for is a long-range comprehensive investment program, conservative yet liberal, which will protect you from the effects of inflation and also deflation, and which will allow you to sleep nights. In this case just stop in at my office and let us recommend a program. I will see to it personally that your inquiries are referred to the Head of our Crystal-Gazing Department.[30]

Conclusions

For a number of years, in a previous life as a tech CEO, I attended the Allen Conference in Sun Valley. At those conferences, I had the opportunity to spend time with Warren Buffett, who would generously share his thoughts with me. I remember asking one year for his opinion on hedge funds and private equity. He paused, shook his head, and said with a smile, "Never have so few been paid so much for so little."

In my view, many of those who perform the important role of setting the price of capital are overcompensated. In large part, this is because individual and institutional investors have been fooled by the winners and fail to correct for survivor bias. I suspect those who run the leading hedge funds would be willing to do so for less than $246,000 per hour.

The extraordinary wages of those in the financial services industry also distort the optimal allocation of another critical resource—some of the nation's most talented workers. The financial services industry draws some of the brightest people away from other professions, such as science, engineering, medicine, and teaching. The extent to which this occurs is difficult to gauge. We will never know how many individuals decided to research stocks rather than cancer. But there are indications that the financial services sector is siphoning off at least some of the best talent: One out of every five students at Harvard goes directly to Wall Street after graduation.[31] Between 20 percent and 30 percent of graduating seniors at the leading Ivy League schools choose finance as their first job.[32]

Spending too much on financial services skews the allocation of resources away from other important, productive sectors of the economy. We would all be better off reallocating a portion of the substantial fees paid to those on Wall Street to other workers. To make this change, we have to first recognize that the financial services industry uses survivor bias to deceive us.

Another group of individuals who have relied upon our failure to correct for survivor bias are the corporate strategy consultants and their brothers-in-arms, the authors of "get rich" schemes. These highly compensated consultants and authors use survivor bias to focus our attention on a small number of winners and ignore the many individuals or companies who were not successful. In fact, this focus on the winners was used to market a book that advised companies on the "real secrets of corporate success" and went on to become one of the best-selling, most well-known business books of all time.

And some firms that followed this advice subsequently filed for bankruptcy.

CHAPTER 2

Strategy and Self-Help Books: Bad Advice

McKinsey's *In Search of Excellence*

In 1982, two McKinsey consultants, Tom Peters and Robert Waterman, wrote *In Search of Excellence*, which became one of the best-selling business books of all time.[1] With more than three million copies sold worldwide, *In Search of Excellence* even has its own acronym—ISOE—within corporate circles. During the 1980s, a number of companies modeled their corporate strategy on the principles laid out in ISOE.

The origins of ISOE began in 1979 when American businesses were struggling. Inflation was over 10 percent, and unemployment had reached similar numbers. Japanese firms were taking market share from American companies by producing superior automobiles and electronics. European firms were also competing head-to-head with American corporations around the world. For the first time since WWII, American companies no longer dominated global markets.

McKinsey saw an opportunity. The firm tasked two separate teams of consultants, one in New York and the other in San Francisco, to identify what makes companies successful. After a year of work, the New York group was unable to reach any conclusions. However, the San Francisco team, headed by Peters and Waterman, believed they had discovered a formula for success, or what they described as the "real secrets of management."[2] The project design of the San Francisco team of Peters and Waterman was that there was no project design. Peters wrote years later that "there was no carefully designed work plan. There was no theory I was out to prove. I went out and talked to genuinely smart, remarkably interesting, first-rate people." He described it as having permission from the company "to talk to as many cool people as I could around the United States and the world."[3]

To select the "cool people" to interview, Peters and Waterman spoke with their fellow McKinsey partners and picked sixty-two companies that were perceived to be the best of the best. The pair then eliminated nineteen of those sixty-two companies because they failed to meet the McKinsey team's test for financial success. To qualify as a financial success, a company had to be in the top half of all companies in at least four of six metrics during the period from 1961 to 1980. The six metrics were asset growth, equity growth, ratio of market value to book value, return on total capital, return on equity, and return on sales. The McKinsey team then interviewed the executives of the forty-three companies that were both "cool" and met McKinsey's test for financial success. Based on the transcripts of the interviews, Peters and Waterman identified eight common characteristics of successful companies.

Each of the eight chapters of ISOE are devoted to one of these characteristics:

1. Bias for action

2. Staying close to the customer

3. Autonomy and entrepreneurship

4. Productivity through people

5. Hands-on, value-driven

6. Stick to the knitting

7. Simple form, lean staff

8. Simultaneous loose-tight properties

The conclusion of ISOE is that firms that wish to achieve excellence and be successful should strive to embody these eight characteristics.

The methodology employed by Peters and Waterman to determine excellence suffers from survivor bias. In their defense, Peters and Waterman never claimed that ISOE was a comprehensive survey of all companies at the time based on statistically valid sampling techniques. Peters has openly stated it was a "subjective nomination process."[4] Regardless, survivor bias ripples through ISOE.

The initial selection of the sixty-two companies was from firms that had survived over the past decades. Furthermore, winnowing the sixty-two companies down to forty-three was based on financial results during the period from 1961 to 1980. However, it is quite possible that the eight characteristics of excellence were also shared by many other companies that were not part of the initial sixty-two or the final forty-three. Many companies with these characteristics of excellence may have been unsuccessful or even gone bankrupt over this period. The fatal flaw in ISOE is that the authors of the study failed to consider whether their eight characteristics of excellence were also typical of firms that were not successful.

To inoculate the study from survivor bias, the McKinsey team should have generated a list of possible characteristics of excellent companies, surveyed all the companies that existed in 1961, and determined the performance of those companies through 1980—including the non-survivors. The team could have then identified the characteristics common to the successful companies. If the underperformers and non-survivors also shared these characteristics, then the conclusions of ISOE would be unfounded. Of course, it is possible that these eight characteristics are the "real secrets of corporate success." However, the methodology of ISOE precludes that determination. In fact, a number of the companies identified by ISOE as exhibiting excellence were not so excellent in subsequent years: Kmart and Wang Labs would file for bankruptcy.

The success of ISOE led others to write similar tomes on corporate strategy. However, a number of subsequent books also failed to correct for survivor bias.

Collins's *Good to Great*

After ISOE, one of the most successful strategy books was *Good to Great*, published in 2001 by Jim Collins. This work promised that "any organization can substantially improve its stature and performance, perhaps even become great, if it conscientiously applies the framework of ideas we've uncovered."[5]

To create their framework, the authors analyzed the strategies of the world's most successful companies—those that had beaten the stock market averages consistently over the past forty years. Eleven companies were singled out from a sample set of over one thousand:

- Abbott Labs

- Circuit City

- Fannie Mae

- Gillette

- Kimberly-Clark

- Kroger

- Nucor

- Philip Morris

- Pitney Bowes

- Walgreens

- Wells Fargo

The authors of *Good to Great* identified seven characteristics common to these eleven companies and asserted that these traits were the keys to corporate success:

1. Leaders who were driven but humble

2. Find the right people

3. Confront brutal facts

4. Define what you are best at

5. Instill a culture of discipline

6. Use technology aggressively

7. Launch small initiatives

The authors of the book stated that they "developed all of the concepts . . . by making empirical deductions directly from the data."[6]

Of course, it is possible the seven characteristics singled out are what makes a company great. But the authors' sample set suffered from survivor bias. First, the authors' sample was composed of companies that had survived until 2001. Companies with the same seven characteristics that had existed forty years earlier but had gone out of business were not part of the data from which the authors made "empirical deductions." Second, the authors failed to consider the companies that had survived over the past forty years and exhibited these seven characteristics but did not overperform the stock market averages. To substantiate their claims, the authors should have determined which other companies shared those seven characteristics forty years earlier but did not survive, and also which other companies shared those characteristics but did not outperform the stock market averages.

In the ten years after *Good to Great* was published, six of the eleven companies identified as "great" underperformed the market.[7] Fannie Mae's stock price sank from $80 per share in 2001 to less than $1 in 2008. Circuit City went bankrupt in 2009.

Hill's *Think and Grow Rich*

A best-selling book from the 1930s was Napoleon Hill's *Think and Grow Rich*.[8] Originally published in 1937, this advice book on how to obtain financial independence has sold more than 100 million copies to date.[9] *Think and Grow Rich* is the twelfth best-selling book of all time, behind *The Da Vinci Code* and ahead of *Harry Potter and the Half Blood Prince*.[10] *Think* and *Grow Rich* is actually based on Hill's previous book, *The Law of Success*, written in 1925, which Hill claimed was penned at the behest of Andrew Carnegie.

The Law of Success was based on evidence collected from Hill's interviews with forty-five millionaires. During the interviews, Hill would ask a series of professional and personal questions to identify the characteristics this group shared in common. He then distilled this information into "the method by which desire for riches can be transmuted into its financial equivalent."[11]

The method has six steps:

1. Fix the amount of money you want.

2. Determine what you will give to get it.

3. Establish a date to get it.

4. Create a plan to get it.

5. Write out the plan.

6. Read the plan twice daily while visualizing the money.

To prove the effectiveness of his methods, Hill cited examples of great business leaders who basically followed this method. He referenced the public statements of Henry Ford, Thomas Edison, the Wright Brothers, and Guglielmo Marconi. He then drew conclusions about what these titans of industry had in common. Two of his conclusions stand out.

The first is that "every great leader, from the dawn of civilization down to the present, was a dreamer." The reason dreaming is the key to wealth is that "if you do not see great riches in your imagination, you will never see them in your bank balance."[12]

The second is that "Man's greatest motivating force is his desire to please a woman . . . men who accumulate large fortunes, and attain great heights of power and fame, do so mainly to satisfy their desire to please women."[13] (Hill's sexist commentary is also an example of how other biases can be layered on top of survivor bias.[14])

Hill goes on to write:

> *Sexual energy is the creative energy of all . . . There never has been, and never will be a great leader, builder or artist lacking in this driving force of sex . . . The desire for sexual expression is by far the strongest and most impelling of all human emotions, and for this very reason this desire, when harnessed and transmuted into action, other than physical expression, may raise on to the status of genius . . . One of America's most able business-men frankly admitted his attractive secretary was responsible for . . . lifting him to heights of creative imagination.[15]*

Putting aside the blatant sexism and whatever this businessman was supposedly imagining, Hill's evidence does not prove out his conclusions. It is possible many of the financially successful individuals Hill wrote about had traits in common. But Hill failed to consider that there may have been many more individuals who demonstrated the same traits but were still dirt poor. Hill based his conclusions on the winners and ignored those who finished behind them in the race to fatten their wallets. He claimed to be laying out a path to financial riches but ignored the fact that others may have followed an identical route to the poor house.

To validate his claims, Hill should have interviewed hundreds or better yet thousands of people, rich and poor, and identified the characteristics unique to the rich. A complete sample would have included people from all economic classes and both sexes, and then he could have potentially identified the characteristics that distinguish the wealthy.

There may have been many individuals who were dreamers, had inappropriate thoughts about their secretaries, and followed the same path as the rich but did not acquire great wealth. Because Hill did not correct for survivor bias, his conclusions were unfounded.

Stanley and Danko's *The Millionaire Next Door*

A similar best-selling advice book on how to reach one's financial goals is *The Millionaire Next Door* by Thomas J. Stanley and William D. Danko. Published in 1996, this book promises to reveal the "surprising secrets of America's wealthy."[16] To uncover these secrets, the authors surveyed thousands of wealthy Americans.

To find the "millionaires next door," more than 300,000 neighborhoods were sorted by zip code, and then within each zip code an average home value was estimated.[17] The basic idea was that the value of a home was a proxy for income. The neighborhoods within a zip code were then ranked from richest to poorest, and an eight-page survey, including a return envelope and a dollar bill, was sent out to randomly selected heads of households in a number of the wealthiest neighborhoods. Three thousand surveys were sent and 1,115 were returned. Of the returned surveys, 385, or 34 percent, reported a net worth of more than one million dollars. The authors used the responses from these millionaires to determine the keys to financial success. The authors claimed that "the research is the most comprehensive ever conducted on who are the wealthy in America."[18]

The authors concluded that the wealthy have seven characteristics in common:[19]

1. They live well below their means.

2. They allocate time, energy, and money efficiently.

3. They believe financial independence is most important.

4. Their parents did not support them financially as adults.

5. Their adult children are economically self-sufficient.

6. They are proficient at targeting market opportunities.

7. They chose the right occupation.

Of course, there is no reason to doubt this sample of wealthy individuals share these traits. However, there may be many poor households that also share these same characteristics. The survey considered only the rich and not those less fortunate.

To validate their claims, the authors should have generated a list of possible factors that might differentiate the wealthy from nonwealthy households and then surveyed a broad section of households across the economic spectrum. If some traits were characteristic only of the wealthy, then their conclusions could be well founded.[20] If both rich and poor households exhibited these characteristics, then the conclusions of *The Millionaire Next Door* are wrong.

Readers were fooled by the winners because the authors of *The Millionaire Next Door* focused only on the wealthy and ignored the less fortunate.

Self-Help Books: Follow the Winner

Survivor bias has also been evident in autobiographical advice books on how to achieve personal success. Many books have been written retelling the life story of a well-known public figure in sports or commerce. Superstar athletes have expounded on the mindset and training practices that enabled them to reach the pinnacle of their sport. Business leaders have revealed their techniques for scurrying up greasy corporate poles. These advice books claim that by following a similar path and practicing similar techniques, the reader can achieve a comparable level of success.

Without question, many CEOs and elite athletes have been remarkably successful based on their talents and hard work. Studying these winners is both instructive and inspirational. However, how a particular individual got to the top is not necessarily the best way to get there.

A particular superstar or business titan could be the exception among many who followed similar paths but were not as successful. A more productive approach would be to study all those who started in a given profession and then to determine the factors common to those who were successful and those who were not. Considering those who did not win as well as those who did is equally important. Steve Jobs, Bill Gates, and Mark Zuckerberg all dropped out of college. But the best path to earning tens of billions of dollars is not to avoid post-secondary education. Most billionaires today went to college, and most people with only a high school education are not billionaires.

To adjust for survivor bias, these self-help books should consider all those who followed the same advice, and then measure their success. Instead, most self-help books measure for success and then look to see what advice they followed.

By failing to correct for survivor bias, many self-help books are not that helpful.

Frank Sulloway's *Born to Rebel*

A similar approach of concentrating on the winners and ignoring those less successful or well-known appeared in the 1997 book by Frank Sulloway, *Born to Rebel*.[21] This best-selling book examined the effect of birth order on personality.

The author argued that firstborn children identify more with authority and therefore are more conservative and conventional. In contrast, non-firstborns question the status quo and hence are rebellious and more adventurous. The author contends that these personality traits are a function of birth order, which affects the family dynamic in which children are raised. Older and stronger firstborns dominate their younger brothers and sisters and thus tend to like the world the way it is. The siblings who follow have a harder time competing with the firstborn and consequently rebel against the status quo, developing a "revolutionary personality." Firstborns are also more introverted and inflexible since they need less support. By contrast, laterborns are more extroverted and agreeable since they need assistance from others to compete with firstborns.

To validate his claims, the author analyzed 121 historical events and the biographical data on 6,500 individuals. For example, Sulloway examined the statements from a number of prominent scientists during the period from 1859 to 1875 when the controversy over Darwinism was raging. The author found laterborns were 4.6 times more likely to support Darwinism than firstborns. He also found that during the same time frame, laterborns were less likely than firstborns to support conservative theories, such as eugenics. After reviewing these historical events, Sulloway expanded his study to present biographical information on various historical individuals. Examples of firstborns were FDR, Churchill, and Stalin; examples of laterborns included Martin Luther King Jr., Malcolm X, and Lenin. (Hitler's home life was complicated:

He was his mother's oldest living son, but there were two older stepchildren in the household.)

The author's survey of 6,500 prominent individuals and more detailed biographical study of the most famous thinkers and political leaders over the last thousand years was obviously not a random sample. The focus on well-known figures in history ignores the billions born into families with more than one child during that same period who did not lead revolutions, either political or intellectual.

In fact, studies have consistently failed to establish a correlation between birth order and personality traits.[22] One study of over twenty thousand randomly selected individuals in the United States, Great Britain, and Germany found "no birth order effects on extraversion, emotional stability, agreeableness, conscientiousness, or imagination . . . we must conclude that birth order does not have a lasting effect on broad personality traits."[23] This research has shown that including the non-survivors disproves Sulloway's theory of the effect of birth order on personality, at least for the 99 percent of us who are not famous (or infamous).

Conclusions

The failure to correct for survivor bias can lead us to implement business strategies that promise success but end up in failure and to follow the bad advice of many advice books. To protect ourselves when drawing conclusions about the keys to success, we should consider the unsuccessful. Considering those who do find success as well as those who don't will stop us from accepting unsubstantiated sweeping claims about birth order and personality, or blindly retracing the educational and professional footsteps of a few famous individuals.

We should not be fooled by the winners. Especially when the winners are the authors of many leading corporate strategy and self-help books.

Next, we'll turn to the world of 1930s publishing when mass-market paperbacks started to become widely available. One of the first authors to exploit this opportunity on a large scale was a botanist who wrote a series of books over the course of four decades that "proved" the ability of some to predict the future.

He called this ability extrasensory perception, or ESP.

CHAPTER 3

Against the Odds: ESP, Lotteries, and Restaurants

Joseph Rhine: 100 to 1

Joseph Banks Rhine (1895–1980) wrote a series of best-selling books about psychic abilities, such as telepathy and clairvoyance. He also invented the term extrasensory perception, or ESP.[1]

Born in Pennsylvania, Rhine was one of five children of a merchant father and schoolteacher mother. His family moved to Ohio when he was in his teens, and there he first became fascinated with psychic phenomena. Later in life, he recalled being mesmerized by magicians who traveled the small towns of the Midwest hypnotizing audience members, placing mothers, fathers, and children in trances. Rhine also had personal experiences with the mysteries of the mind. As a boy, he would regularly sleepwalk: His parents once found him half a mile down the road. Other times he would talk at length with others while asleep, standing with eyes shut and arms dangling at his sides, yet remembering nothing of the conversation the next morning. He also endured an attack of the measles at nineteen that left him with little sense of smell or taste and near blindness in one eye for the rest of his life.

After graduation from high school, he enrolled at The College of Wooster to study religion. However, he lost interest in school and in 1918 joined the Marines. In an example of the military's sometimes perverse sense of humor, he was trained to be a sharpshooter. While in the service, his beliefs swung decidedly against all forms of religion, and he resolved to become a scientist. He enrolled at the University of Chicago and earned a PhD in botany in 1925.

During Rhine's studies at the University of Chicago, his interest turned to the paranormal. He attended a lecture by the author of the Sherlock Holmes stories, Sir Arthur Conan Doyle, who espoused spiritualism. A gifted and charismatic speaker, Doyle regaled his audience with tales of conversations with his deceased brother and appearances in ghostly form of his equally dead mother. Doyle told the spellbound audience, "I swear by all that's holy on earth, I looked into her eyes." Doyle stated flatly that spiritualism was of "transcendental importance" and it would be "unscientific to ignore it."[2]

After graduation from the University of Chicago, Rhine joined the Psychology Department at Harvard with a goal of establishing parapsychology as a new branch of the discipline. He left Harvard in 1927 to start up a parapsychology institute at Duke University and remained there for more than five decades.

At Duke, Rhine set out to gain the respect of the scientific community and general public for his work in parapsychology. His first book, *Extra-Sensory Perception*, published in 1934, tells of his discovery of eight individuals with ESP: A. J. Linzmayer, a Duke undergraduate; Charles Stuart, one of his research assistants; Hubert Pearce, a Duke divinity student; and five other unnamed Duke students. His work led him to conclude "that ESP is an actual and demonstrable occurrence" and that it happens under "pure-clairvoyance conditions, with not only the sensory and rational functions, but telepathic ability as well."[3]

Unlike others before him, Rhine employed objective standards to test paranormal activity. Although he refined his methods over time, the basic test Rhine employed remained roughly the same. A deck of twenty-five cards, each with one of five signs—a circle, a rectangle, a star, wavy lines, and a plus sign—was shuffled and the subject asked to identify the image on the back of the card. Five correct guesses in a row were considered statistically significant. In the back of his 1957 book *Parapsychology*, Rhine supplies a table that sets out the number of correct answers needed to prove a subject was gifted with ESP.[4] Based on this table, Rhine claimed that "the odds are 100 to 1 or higher" that the evidence of ESP he obtained was due to chance.[5]

In fact, it was survivor bias at play.

Imagine a questionnaire is sent to thousands of individuals asking them to predict the outcome of a horse race. Those who predicted the first race correctly are sent another questionnaire to forecast the results of the next race. Those who were right a second time are surveyed again, and so on. After a number of rounds, it is likely that at least several individuals will have called every race correctly. Almost certainly, some survivors of this process will have demonstrated what seem to be remarkable skills at predicting winners. If an experimenter tests enough subjects, several people are likely to survive the process. These "winners" will appear to be exceptionally skillful at placing wagers at the track.

To find eight individuals who appeared to possess ESP, Rhine conducted tests on thousands of students over several years. Given the large number of individuals tested, he was highly likely to find eight individuals who were able to consistently and correctly guess many of the images on the backs of cards.

In addition, it has been alleged that Rhine "stacked" the deck. Some claim that the cards employed were particularly thin and the images slightly perceptible in the right light. Also, many of the interviewers wore glasses, and the images on the cards may have been reflected in their lenses. The subjects of Rhine's tests may not have been consciously cheating, but the validity of his methods have been criticized. Regardless of whether Rhine's findings were the result of survivor bias or just flawed experiments, subsequent attempts to replicate his findings have been unsuccessful, including tests on some of the eight individuals Rhine asserted exhibited ESP.

Based on his experiments, Rhine published a series of popular books. His first book on ESP in 1934 was followed up with books in 1937, 1940, 1947, 1953, 1957, 1965, 1968, and 1971. From the 1930s through the 1970s, he was regularly featured in magazines and appeared on radio and television. Rhine used his public appearances and books to encourage millions of Americans to conduct experiments at home to discover if friends and family exhibited signs of ESP. To aid the experiments, Rhine marketed his own brand of cards printed with the five images from his experiments at Duke. His fame spread far and wide. Richard Nixon, Aldous Huxley, Carl Jung, Jackie Gleason, and Albert Einstein visited Rhine at his lab. In 1961, Timothy Leary convinced Rhine to experiment with LSD in order to determine whether the drug could induce paranormal activity. Rhine reported it didn't.[6]

Rhine fooled us by heavily promoting the handful of "winners" of his ESP experiments while making no mention of the thousands who had "lost." Rhine convinced millions that ESP was real and they should use his playing cards to "test" their family and friends for telepathic gifts.

Those who promote ESP to sell books and other items are not the only ones to profit from our tendency to be fooled by the winners.

Lotteries: A Dollar and a Dream

In 2018, Americans spent $77 billion on lotteries.[7] Forty-five states offer their residents a chance to strike it rich. To encourage ticket sales, lotteries aggressively advertise the jackpots won by a few lucky individuals. I am unaware of a TV commercial that features a poor family skimping on meals after overspending on lottery tickets.

Unfortunately, in most states, the purchasers of lottery tickets are disproportionally nonwhite, urban, male, and on government assistance.[8] Hence, lotteries are a form of regressive taxation. One study showed that households that make less than $12,400 a year spend 5 percent of their income on lotteries.[9] Overspending on lotteries can tragically become a vicious circle: The more spent on lottery tickets, the worse a family's financial situation and the more likely they are to buy more lottery tickets as a desperate measure to alleviate financial distress. In terms of public policy, the best that can be said about lotteries is that at least it is a voluntary form of taxation.

Lotteries have been called "a tax on those who can't do math." This is because the payouts, or the percentage of ticket sales allocated to jackpots, range from 20 to 80 percent, depending on the state and the particular drawing.[10] The purchase of a lottery ticket almost always has a negative expected financial return. But not for states. Eleven states collect more monies from lotteries than from corporate taxes.[11]

The proponents of lotteries argue that they are a form of entertainment. Players may spend hours before a drawing imagining life after winning. One state lottery has the motto, "All you need is a dollar and a dream." In that sense, what lotteries are really selling is hope. Nearly every ticket purchaser knows lotteries are not a good financial bet. Nevertheless, purchasers of lottery tickets are willing to fork over hard-earned money, week after week, if only to daydream for a bit about sand, sun, and sea. But that does not explain why so many buy multiple tickets. The odds of winning a large jackpot in most cases are one in many hundreds of millions. It is not clear why purchasing five tickets rather than one will increase the enjoyment that comes from imagining a better life.

If states wanted to offer "hope" as a public good, then lotteries could be set up that are more citizen friendly. For example, a state could run a lottery that paid out 100 percent of ticket sales and limited purchases to one ticket per person. In such a lottery, individuals could wager a token amount of money with a net zero expected value, and then fantasies about luxury yachts and umbrella drinks are a freebie.

Restaurants: Not Always Open

Lotteries are an example in which focusing on the winners leads us to overspend on tickets. This particularly affects many poor families, who can least afford this form of entertainment. Opening a restaurant is a similar but less extreme example. For many workers stuck in low-paying, dead-end jobs, restaurants give the appearance of providing a path to financial independence with the added benefits of working indoors and the chance to be your own boss.

But many who open dining establishments are overly optimistic about the odds of making their personal and financial dreams come true. Their exuberance comes from being fooled by the winners.

The failure rates of restaurants are high—about 70 percent are out of business or under new ownership within three years after opening.[12] Yet, it is understandable why prospective new restaurant owners may overestimate their chances for success. In every town I have lived in, many restaurants seem to have been around for a long time, in some cases decades. This could lead one to believe that restaurants are a relatively reliable, stable business.

But a number of studies have shown that the failure rates for restaurants follow a particular pattern. One study showed that the failure rate during the first year of operations is 26 percent and remains elevated for years, before declining to a low single-digit number by year seven.[13] Hence, prospective restaurant owners walking about their local towns mostly pass by the surviving restaurants, the relatively small minority that made it through the first several years, until the failure rates dramatically decline. If the non-survivors, all the restaurants that failed, were left empty, this would scare off many prospective restaurateurs.

Local governments could easily provide an accessible public listing of eating establishments that have opened and closed, as restaurants have to obtain and renew business licenses and food handling permits. I suspect fewer individuals would quit their jobs and open a restaurant, only to soon fail, if the non-surviving establishments were more readily observable.

Conclusions

Joseph Rhine was one of the best-selling nonfiction authors of his time, one of the first to exploit the new mass market for paperbacks, as well as a successful marketer of his branded ESP cards. Rhine used survivor bias to deceive us by focusing our attention on the few "winners" of his experiments and failing to point out the thousands of "losers" who demonstrated no evidence of ESP. Similarly, states heavily promote the "winners" of lotteries to generate tax revenues from those who can least afford it.

In the food services industry, highly visible "winners" and equally invisible "losers" lead those looking to open a restaurant to be overly optimistic about their chances for success. Almost all restaurants would be completely empty if those that had failed were still open.

The food industry is not unique in the lack of accessible public information that tracks failures as well as successes. In the health care industry, medical journals rarely publish studies with negative outcomes. The research that proves that a drug or surgical procedure is ineffective most often ends up permanently filed in some laboratory drawer.

The first person to formally identify how we are fooled by the winners in medicine was actually a computer scientist. He pointed out that evidence indicating that a drug or medical procedure does not work is often ignored and, as a consequence, we spend monies on health care that does not make us healthier. But his purpose in calling out survivor bias in medicine was not to reduce medical bills.

He used survivor bias to deceive many into believing that smoking doesn't cause cancer.

CHAPTER 4

Medicine:
Unhealthy Confidence

Theodor Sterling: Thank You for Smoking[1]

Theodor Sterling (1923–2005) was born and raised in Vienna. His family moved to America in 1940 when he was seventeen years old, and he soon thereafter enlisted in the US Army. After the war, he attended the University of Chicago and subsequently received a PhD in computer science from Tulane University. He taught at several universities in the United States and moved to Canada in 1972, joining the Computer Science Department at Simon Fraser University in Vancouver. In 2001, Simon Fraser awarded him an honorary doctorate of science. Before passing away in 2005, Sterling endowed the Simon Fraser University Sterling Prize in Support of Controversy, a $5,000 grant given each year to "honor and encourage work which provokes, and/or contributes to the understanding of controversy."[2]

In a paper published in 1959, Sterling was the first to formally identify survivor bias in medical journals. He wrote, "There is some evidence that in fields where statistical tests of significance are commonly used, research which yields nonsignificant results is not published."[3]

This problem is still prevalent today. Here's an excerpt from a typical letter sent by the editor of a major medical journal explaining why a paper that detailed why a drug didn't work was rejected for publication: "The manuscript is very well written and the study well documented. Unfortunately, the negative results translate into a minimal contribution to the field."[4]

Decades ago, Sterling pointed out that medical research validating a positive result is regularly published, but studies that yield a negative outcome rarely see the light of day. For example, Sterling observed that of one hundred randomly selected papers published in a prominent medical journal, fully ninety-four documented positive results and only five documented negative outcomes (the remaining study reproduced a previous study's results).[5]

The reasons are that editors regularly reject submissions with negative results, and researchers infrequently submit results that demonstrate a drug or procedure is ineffective because they know such papers will be rejected. In effect, medical researchers have learned to self-censor. Studies have shown that "the main reason investigators fail to submit their research for publication are the negative results themselves."[6] As a result, clinical trials that conclude a drug or medical procedure is effective are the ones most frequently submitted and published. When physicians and government officials open the pages of medical journals, almost all of the articles they see are reporting positive outcomes.

Sterling argued that in order for us to make smart decisions about our health care we need to review all clinical trials, including those with negative results.

In addition to the bias shown by editors of medical journals, drug companies have compounded the problem. Drug companies, who fund and control the publication rights for many clinical trials, consistently block disclosure of negative outcomes. In a survey of two thousand oncology trials, drug companies published results only 6 percent of the time, reporting positive outcomes in three-fourths of the published studies.[7] The same survey showed that clinical trials funded by hospitals or clinics published studies about 60 percent of the time and reported positive results in only about half of all outcomes.[8] Because studies with positive outcomes are much more likely to be published, we pay for drugs and undergo medical procedures that do not make us healthier.

Sixty years after Sterling first identified survivor bias in medical studies, survivor bias continues to undermine our health care system. A recent survey of 4,600 medical publications showed that survivor bias has been increasing over time.[9] Today, roughly half of completed clinical trials still go unreported.[10] A review of the US health system concluded that the consequences of this selective publication of outcomes are "medical treatments that are unnecessarily dangerous, ineffective, unpleasant and costly."[11]

Sterling rightly called out the drug companies and medical journals for failing to publish studies that yielded negative results—the non-survivors of medical research. But he had a particular motivation for doing so.

Sterling, a computer programmer by training, was paid millions in consulting fees as a "medical scientist" by the tobacco industry over the course of his career. As a consultant, Sterling spent a large part of his professional life trying to convince others that there was no scientific proof that smoking causes cancer. He argued that many medical studies that could not identify a direct connection between lung cancer and smoking were ignored. He claimed that the negative results of these studies should be weighed equally by doctors and health officials against the studies that proved a causal link between lung cancer and the use of tobacco.

Sterling received most of his consulting fees from the Council on Tobacco Research (CTR). This research arm of the tobacco industry was founded in 1958 and remained operational until it was dissolved in 1998 as part of the Tobacco Master Settlement Agreement. The CTR led the fight to persuade Americans and the rest of the world that cigarettes are not harmful to human health. A memo written in 1969 from the head of marketing for Brown & Williamson outlined the objective of the CTR: "Doubt is our product since it is the best means of competing with the body of fact."[12]

During the 1960s and 1970s, CTR concentrated much of its research monies on nicotine, because nicotine doesn't cause cancer. This produced a tall stack of papers detailing clinical trials that showed no tie between nicotine and lung cancer. Sterling then argued that these negative results were overlooked by the doctors and public health authorities. However, during the 1980s, it became clear that nicotine was addictive, and so CTR stopped funding nicotine research.[13]

CTR then turned its efforts to finding other ways to muddy the research waters. The CTR attacked the scientific consensus that smoking causes cancer through an initiative within CTR known as "Special Projects."[14] In all, 140 special projects were funded by CTR.

Here are three examples:

- Special Project 68: Emphysema could be due to an initial insult by virus or another organism.

- Special Project 98: Environmental factors cause both smoking and heart disease, thus generating a correlation where no causal relationship exists.

- Special Project 112: Lung cancer detection bias helps to perpetuate the belief that cigarette smoking is the main cause of lung cancer.

Sterling was called "CTR's most trusted ally."[15] From 1968 to 1990, Sterling received more than five and a half million dollars for completing Special Projects.[16] His first assignment was to review a surgeon general's report that linked smoking with cancer as well as heart disease.[17] He was also tasked with refuting National Health Survey data that linked secondhand smoke to cancer deaths. In response to these reports, he blamed bad air in office buildings and cities for lung cancer. He cited factors such as "lighting and ventilation, type of air, [and] presence of smog types."[18]

In 1978, Sterling published a paper in the *International Journal of Health Services* titled "Does smoking kill workers or working kill smokers?"[19] Just as he argued bad air may be responsible for lung cancer, Sterling claimed in this article that "a person's occupation, not cigarette smoking, may be the primary cause of lung disease, especially of cancer and chronic obstructive disease."[20] Sterling cited data that showed blue collar workers suffer greater exposure to toxic fumes than white collar workers and therefore exhibited higher rates of lung disease. Based on this data, he concluded that the sicknesses attributed to smoking may, in fact, "be due to occupational exposure rather than personal habits."[21]

Sterling regularly testified before Congress. In his opening statement to the House Committee on Interstate and Foreign Commerce, Sterling declared:

> *Information collected so far on the link between smoking and various diseases has raised a certain amount of disagreement in the scientific community . . . a number of animal studies have*

*attempted to produce lung cancer by means of
cigarette smoke. The results of these studies have
been negative.*[22]

Sterling then proceeded to assert that it is smokers' carpe diem
attitude that causes lung cancer:

> *The association one observes may not be between
> cigarette smoking and disease, but it may be between
> a certain style of life and disease . . . heavy smok-
> ers also drink alcoholic beverages excessively . . .
> smokers are said to be more easily angered . . . they
> have been reported to marry oftener, to change
> jobs more frequently . . . the general picture emerg-
> ing from all studies is one of smokers tending to
> live faster and more intensely.*[23]

In his testimony, Sterling went on to criticize the scientific
community for ignoring the numerous studies of smoking and
lung cancer that generated negative results, echoing many of the
concerns he raised in his 1959 paper about survivor bias in medical research.

Sterling was wrong about smoking and cancer, but he was
right about drug companies and medical journals. He correctly
called out those who trumpet only the positive outcomes of clinical trials. Regardless of his motivations, Sterling deserves credit
for his efforts to call out those in medicine who use survivor bias
to deceive us.

Modern Drug Studies: They Probably Don't Work

Sixty years later, survivor bias continues to plague the health
care industry.

One of the most successful group of drugs in recent history has been selective serotonin reuptake inhibitors (SSRIs). These antidepressants have been extremely popular, constituting five of the thirty-five most widely prescribed drugs in the United States.[24] Marketed under the brand names of Cymbalta, Effexor XR, Lexapro, Wellbutrin, and Zoloft, SSRIs are taken by hundreds of millions of patients each year, raking in billions of dollars of revenues for drug companies.[25]

A study of 74 trials of SSRIs with over twelve thousand patients showed 38 positive and 36 negative results. All but one of the positive trials were published.[26] Of the 36 negative trials, 22 were never published.[27] The authors of this meta-study of SSRI trials observed that this bias against reporting negative results is a substantial issue and warned that "antidepressants are a prime example of over-medicalization of our society."[28] The authors concluded the study with the following statement: "We are blinded by a blizzard of small, selectively designed, analyzed and reported trials." The authors go on to say that, although this may be depressing news, "Nevertheless, even if one feels a bit depressed by this state of affairs, there is no reason to take antidepressants, they probably won't work."[29]

Sometimes research is not published in order to avoid disclosing adverse reactions. A group of former drug researchers revealed the results of five clinical trials of an SSRI specifically formulated for treatment of depression in children.[30] The drug company that funded the trials allowed only the one study with a positive outcome to be published and none of the four that showed the medicine was ineffective. Most worryingly, some of the unpublished studies indicated an increased risk of suicide.[31]

A similarly troubling pattern can also be seen in a study of an anti-arrhythmic drug tested on ninety-five patients who suffered from myocardial infarction. To gauge the drug's effectiveness, the group of patients was split: Half were given treatment and the other half a placebo. Of the patients given treatment, nine died compared with one on the placebo. The study was not made public until thirteen years later.[32]

Recently, a few doctors have begun to speak out. In 2016, a research team publicly disclosed survivor bias in their lab's research on oxytocin.[33] Oxytocin has been marketed as an aphrodisiac and has been called the "love hormone." The drug is equivalent to chemicals produced in the hypothalamus, which are transported to the pituitary gland at the base of the brain for release into the bloodstream. This release of these chemicals by the body is associated with childbirth and lactation. It also appears to be related to erections in men and orgasms in women. One study found that oxytocin in the bloodstream rises when petting a dog.[34]

This research team had performed nine other studies on oxytocin from 2009 to 2014 on a total of 453 subjects. In these studies, oxytocin was used as a nasal spray to attract the opposite sex. Nine articles were written as a result of the research and submitted to medical journals. The four articles that showed positive results were accepted, but just one of the five that demonstrated a negative outcome was published. The other four articles with negative outcomes were rejected despite repeated submissions to multiple medical journals.

After this experience, the research team decided to speak out because "our publication portfolio has become less and less representative of our actual findings." They were concerned about the large number of unpublished studies that found that the drug was ineffective. The authors noted that their enthusiasm for intranasal oxytocin "has slowly faded away over the years and the studies have turned us from believers into skeptics."[35] The team concluded by recommending that researchers "get these unpublished studies out of our drawers and encourage other laboratories to do the same."[36]

Surgeries and Trauma Care: Not Dead upon Arrival

Another area in which the deceptive influence of survivor bias crops up in medicine is in studies on the effectiveness of surgical procedures. A number of studies have shown that, after admission to the hospital, those who undergo surgery live longer than those who don't. Such studies are typically used to validate that a particular surgery is an effective treatment and often cited by those who perform many surgeries.

A review of one of these studies on patients with endocarditis, an inflammation of the heart's inner lining, was undertaken to determine whether there were other reasons the non-survivors, those not operated upon, died earlier and more often.[37] It turns out that many non-survivors were so sick upon admission that they could not be safely operated on. In other cases, patients were so ill that they died after admission but before lifesaving surgery could be performed. These two factors accounted for the differences in mortality rates between those who had surgery and those who didn't. The study concluded that "survivor bias significantly affects the evaluation of surgical outcomes."[38]

Survivor bias has also been demonstrated in studies of trauma care. About 80 to 85 percent of people in North America live within a one-hour drive of a trauma center, but 30 to 60 percent of trauma patients are transported first to the nearest hospital.[39] Researchers studied 11,398 severely injured patients who ended up in trauma centers. About two-thirds were rushed directly to a trauma center, and the other one-third were hurried to a nearby hospital and then later transferred to a trauma center. In total, 2,065 people died, and the survival rates for both groups were about the same. Hence, diverting a third of the severely injured patients to non–trauma centers did not seem to affect patient outcomes. This was good news for local hospitals.

However, it was discovered that about half the patients sent first to non–trauma centers died more than 2.5 hours after arriving at the facility, and this accounted for 22 percent of all deaths.[40] Most of those patients could have made it to a trauma center and thus increased their chances of surviving. If the non-survivors from the non–trauma centers, who died more than 2.5 hours after admission, had been included in the mortality figures, then the study would have reached the opposite conclusion.

Turns out, it was worth the extra miles—if you want to live.

Publishing Negative Results: Learning from Losers

In areas outside of medical research, a few individuals and institutions have set up journals specifically to publish negative results. In 2019, the economics journal *Series of Unsurprising Results in Economics (SURE)* was established.[41] *SURE* exists to find results such as "this intervention does not have any effects" or "there does not seem to be a strong relationship among any of these variables."[42] *SURE* is an online-only, open-access, no-fee journal that seeks papers with negative outcomes that were rejected by other economics journals. One of the first papers *SURE* published showed that spending time informing students about the benefits of taking more credit hours did not prompt students to take more classes or graduate sooner. This is an example of a negative outcome that can be quite useful: Counselors now know not to spend time trying to convince students to load up on more classes.

A remedy for survivor bias in the medical field would be for the government to fund the equivalent of a *SURE* for clinical trials in medicine. Such an effort would not be cheap. It would require a large staff of editors and reviewers to ensure the publication was not flooded with poor quality papers. However, a listing of all validated trials, whether successful or not, would be a counterweight to the survivor bias that burdens medical journals today. This complete data set—the results from all clinical trials, regardless of the outcome—would enable medical professionals and government health officials to make better judgments about the effectiveness of medicines and medical procedures. By including the studies with negative outcomes, government health officials could deny reimbursement for drugs and medical treatments that most researchers believed were ineffective.

Equally important, the government should require drug companies to publish the results of all clinical trials. The drug companies should no longer be able to conduct ten studies, publish the one with a positive outcome, file the other nine in a lab drawer, and market a drug that, while safe, is not generally effective.

Failing to adjust for survivor bias in medical research is one reason many countries each year are spending more on health care than ever and yet are achieving no better outcomes. An important first step on the road to reducing medical costs would be for medical professionals and public health officials to place equal weight on clinical trials that yield negative and positive outcomes.

The data to correct for survivor bias in medicine is out there. We have complete data sets on the number of positive and negative outcomes—the survivors and non-survivors of drug trials and medical procedures. We should not be fooled by the "winners" in clinical trials.

Weight Loss: Fooled by the Few

Another health care–related business in which we are deceived by survivor bias is the weight loss industry. We spend billions on books, diets, and weight loss programs that don't help us shed unwanted pounds.

The weight loss industry's revenues come primarily through the sale of paperbacks and weight loss programs. Companies promote the sale of these products by spending large amounts of advertising dollars on television and social media. Nutrisystem splurges on over $300 million per year in ads.[43] In a recent year, the advertising budgets for Weight Watchers and Jenny Craig consumed $117 million and $34 million, respectively, of company revenues.[44] In 2018, the commercial weight loss companies served up more than $3 billion in sales.[45]

Yet, many weight loss ads have been found to be deceptive. A Federal Trade Commission (FTC) report determined that 55 percent of all weight loss ads were probably false.[46] As an example, the FTC cited a TV commercial for a product composed of ground-up shells of shrimps, crabs, and lobsters, which was promoted with statements such as: "Have you ever seen an overweight fish? Or an oyster with a few pounds too many?"[47]

A common promise made by weight loss companies is that by adopting a particular diet or meal plan, the purchaser can quickly and easily attain a slimmer figure. Offered as proof is a testimonial from an individual who "lost 27 pounds in 27 days," complete with convincing before and after pictures. No doubt this was the case for that individual, otherwise FTC lawyers would come calling regarding truth in advertising.

Whether or not one individual lost weight, however, is not convincing evidence that the diet works for most. What we want to know is how many individuals started on a diet and how much weight the average individual lost. If hundreds of individuals started on a diet, and the only individual who lost weight was the guy on the TV, we should be skeptical of the diet's efficacy. Testimonials alone do not tell us much useful information, as those appearing in the ads have been carefully selected by the weight loss company from among all dieters. A more compelling ad would feature a randomly selected sample of individuals from those who began the diet, or better yet one that disclosed the average percentage of body weight lost by all dieters. These late-night TV testimonials suffer from survivor bias in an attempt by the weight loss firm to mislead the well-served viewer.

The Atkins Diet: 45 Million and Still Growing

One of the most widely followed weight loss programs over the last half century has been the Atkins diet.

Developed by Robert Atkins, a medical doctor, this program recommends a meal plan based on the premise that a low-carbohydrate diet burns more calories than other diets. In practice, this means eating meat and high-fat dairy products at the expense of grains and sweets to gain a "metabolic advantage." Dr. Atkins claimed that steak and eggs increase the rate at which you burn calories when compared to the rate at which you burn calories from bread, rice, and processed snacks. He rightly points out that the most significant change in the human diet has been the remarkable increase in the use of sugar. As little as several hundred years ago, the average American consumed four pounds of sugar annually; today that number is over 175 pounds per year.[48] Dr. Atkins correctly labeled sugar the "killer carbohydrate."[49]

As proof that his method works, Dr. Atkins wrote a series of books packed with examples of individuals who lost weight by consuming large quantities of steak and eggs. One example is that of an author and newspaper columnist, Doris Lilly, an early adopter of his diet.

She reported:

> I was struggling into a size 16 and there aren't any size 18s except in the tent department . . . then one night I was on the Merv Griffin Show, wearing a new shiny Norell dress. It was a taped show. When I saw myself on the screen, I cried! I looked like a silver tub.[50]

But then she was introduced to Dr. Atkins:

> *I went to him and lost forty pounds, twenty of them the first month . . . All it takes is guts. You certainly get plenty to eat . . . I now wear a size ten . . . even my feet are smaller. I gave away all my shoes, in fact all my clothes but my handbags and fur coats.*[51]

There is no reason to doubt that Ms. Lilly was truthfully reporting her experiences on the Atkins diet. Nor is there reason to doubt the other dozens of examples of individuals who lost weight after adopting the Atkins meal plan and whose tales fill up hundreds of pages in Dr. Atkins's books. However, that does not mean the Atkins diet is an effective way for most people to lose weight. Studies have shown over the course of repeated trials that those on the Atkins diet lost an insignificant percentage of body weight.[52] Yet books about the Atkins diet have sold over 45 million copies, and the number continues to grow.

Other popular diets are no different. Controlled studies of commercial weight loss programs have demonstrated that most individuals lose little or no weight. A meta-review of 141 commercial and proprietary weight loss programs and thirty-nine randomized controlled trials lasting from twelve weeks to twelve months showed these programs yielded an average weight loss that was only a small percentage of total body mass.[53]

Dr. Atkins and other purveyors of weight loss books and programs have created a multibillion-dollar industry based on using survivor bias to deceive us. We are fooled by the winners, those whose before and after pictures are so convincing.

Cats Falling from Tall Buildings

We have discussed how survivor bias directly affects human health. But survivor bias also impacts the life and death of some of our favorite animals.

Cats are arboreal creatures. In the wild, the distant ancestors of today's domesticated cats spent much of their lives in trees to protect themselves from predators and to pounce on prey from above. As a result, cats are naturally attracted to open windows in tall buildings and enjoy perching on sills, peering down at the streets below. Not surprisingly, natural selection has bred in cats a remarkable ability, not seen in most animals, to fall from trees and survive. It is still useful today for a cat who resides in a high-rise apartment and whose human occupants leave the windows open during warm weather.

Fortunately, the trees in Africa are generally not fifty stories high, and anyway cats are wise enough not to climb to the top of those that are. Unfortunately, many high-rise apartment buildings have heights that exceed that of the average African tree, and cats have limited influence over where they reside.

An oft-cited 1987 study of cats taken to pet emergency rooms after falls from high-rise apartment buildings discovered a peculiar pattern. As expected, cats that fell from heights of five stories or less were more likely to survive than cats that fell from heights of five to nine stories.[54] Surprisingly, however, above nine stories the mortality rates for cats declined. In short, cats that fell from heights between five and nine stories were the most likely to perish.

Physicists have actually taken time away from studying subjects like quantum mechanics to research the mortality rates of cats falling from high-rise apartment buildings. Of the many models I have studied, the most comprehensive and detailed analysis was published by a physicist in 2017, complete with a mathematical model that predicted the survivability factor as a function of the height of the fall.[55] In this model, the survivability factor is determined by two variables: speed of impact and apparent weight as a function of height.

A higher speed of impact decreases the odds of survival. Until terminal velocity is reached, the severity of injuries correlates directly with the height of the fall—the greater the height of the fall, the more severe the injuries. After reaching terminal velocity, the severity of injury, all else equal, should remain constant. The good news for cats is that natural selection has gifted them with an unusually high ratio of underbody surface area to weight. This means that cats reach terminal velocity in a shorter period of time than most other similarly sized animals. A high ratio of surface area to weight makes evolutionary sense: Cats spent millions of years in trees, and cats that survived falls from trees had more kittens. (This is a reason dogs are less likely to survive falls from heights. The ancestors of today's domesticated dogs were wolves who lived and hunted on the ground.)

In contrast to the speed of impact, a higher apparent weight increases the odds of survival. Apparent weight is the magnitude of force acting against gravitational force, or what the cat perceives as the force of gravity on its body. When a cat first falls, it feels weightless, and its natural instincts take over; the cat rotates its feet underneath its body to brace for impact. For falls from lower heights, this works just fine. However, for falls from greater heights, landing paws down is dangerous: A cat's feet are a relatively small surface area over which to absorb the force of the fall. Thus, for falls from greater heights, it is better to land feet up or sideways to distribute the shock from impact across a broader area of the body.

The average cat reaches terminal velocity at approximately ten stories. At this point, the force of air resistance equals the force of gravity. The cat no longer feels weightless, and the instinctive reaction to land feet first goes away. Some cats still land on their feet after reaching terminal velocity, but most hit the street on their sides or back, spreading out the force of the impact over larger areas of their body. This decreases the severity of injuries and the incidence of mortality.

The model developed by the physicist who studied falling cats combined these two factors to estimate the severity of life-threatening injuries as a function of the height of the fall. For falls from four stories or less, the speed of impact is lower so cats can land paws down and frequently survive without life-threatening injuries. For falls between five and nine stories, cats will not have reached terminal velocity, and apparent weight will not have yet normalized. Unfortunately, at these heights, most cats land paws down and at a higher vertical velocity compared with falls of fewer than five stories. For falls greater than nine stories, cats approach terminal velocity, apparent weight normalizes, and they will often rotate from paws down to paws out or up. By taking into account both speed of impact and apparent weight, this model correctly predicts that the greatest risks to cats are falls from five to nine stories, consistent with the empirical evidence concerning the fatality rates of cats that fall from tall buildings.

I am not a physicist, but this all seems perfectly reasonable to me. However, alternative explanations, related to survivor bias, have also been put forward.

The most common explanation is that most cats that fall from nine stories or more die on impact and therefore are not transported to a pet hospital. This would seem about as clear-cut a case of survivor bias as it gets.

I do not find this convincing. Cats that die on impact and are not taken to a pet hospital are not part of any sample included in the studies of injured cats. The number of deceased cats should not impact the survival rates of cats transported to a pet hospital, as mortality is calculated in all known studies on cats post-admission.

My own theory is that survivor bias does play a role, but it is due to uninformed cat owners.

The physics of cats falling from tall buildings suggests that mortality rates should be positively correlated with height until the terminal velocity is reached and apparent weight is normalized. However, many cat owners are probably not well versed on the physics of cats falling from tall buildings. The intuition of most cat owners is probably that the greater the height of the fall, the lower the likelihood that the cat will survive. So, cat owners mistakenly conclude that for higher falls the risk/reward ratio of rushing their cat to an animal hospital is not worth it. Hence, survival rates increase with the height of the fall after nine stories because the only cats rushed to the emergency room after such falls are the few not obviously seriously injured or deceased.

I cannot prove this. Nor am I suggesting further research should be done.

Conclusions

Theodor Sterling used survivor bias to deceive us about the link between smoking and cancer. While Sterling was wrong to "muddy the waters" for tobacco companies, he was right to shine a light on the deceptive publication practices of drug companies. Sterling and others subsequently have tried to show that we could learn a lot from the "losers" of clinical trials in medicine. Similarly, the weight loss companies have used survivor bias to deceive us into believing that following their particular diet plan is an easy, effective way to shed unwanted pounds, despite all evidence to the contrary. TV ads focus all our attention on the differences between the before and after photos of a few and fail to present the photos of the many in which there has been no discernable change in waist size.

The drug and weight loss companies have an incentive to bury the information about the negative outcomes from clinical trials of medicines and diets. Forcing those industries to publish all outcomes, positive and negative, would enable consumers to spend their monies on medicines and diets that actually work.

Next, we turn to survivor bias in warfare. We begin with the most often cited and most famous example of survivor bias, the story of a mathematician working for the US government during WWII. This former professor had fled Nazi-occupied Europe before he was recruited for a top-secret research unit working out of a series of unmarked rooms in an apartment building near Columbia University.

What he discovered was that US military commanders had been fooled by the winners, by the planes that survived bombing runs over Germany and safely returned to base. The mathematician's analysis and recommendations would go on to save the lives of US airmen over the skies of Germany and Japan, and later in Korea and Vietnam in the decades to come.

CHAPTER 5

Warfare: Bombers, Helmets, and Tourniquets

Abraham Wald: The Missing Bombers

Abraham Wald (1902–1950) was born in Klausenberg, Germany, then a part of the Austro-Hungarian Empire.[1] His grandfather was a famous rabbi, and his father was an Orthodox Jewish baker. One of five children, Wald demonstrated at a young age an unusual aptitude for mathematics. However, as a descendant of Orthodox Jews, he was denied admission to the local academies and was homeschooled by his parents. After attending King Ferdinand University, he entered graduate school at the University of Vienna, earning a PhD in mathematics in 1931. But his Jewish heritage made an academic post in Austria after graduation impossible, so he took a job at the Austrian Institute for Economic Research.

In 1938, he was fired by the newly installed Nazi director of the Institute and left Europe for America. His decision to leave Europe was prescient. Eight members of his family—his parents, sisters, and all but one brother—would later be killed at Auschwitz.

His first job in the United States was as a fellow at the Cowles Commission, an economic institute in Colorado, but soon he moved to New York to teach mathematics at Columbia University. His greatest academic achievement was the discovery of the theory of statistical decision functions. Today, Wald is known as the founder of sequential analysis. He also published some of the first papers on game theory.[2] The methods of sequential analysis he developed were widely applied to the US WWII effort.[3]

Wald was known to his colleagues to be completely immersed in his work, with little interest in anything else other than his family. He preferred to think while walking, sometimes working out his papers while talking with colleagues on long hikes in Morningside Park near Columbia. Wald was also good friends with Kurt Gödel and John von Neumann and periodically traveled to Princeton to discuss mathematical discoveries deep into the night at their homes. He married in 1941 and had two children, Betty and Robert (the latter is currently a well-known physicist at the University of Chicago). In 1946, Wald founded the Department of Mathematical Statistics at Columbia, and his lectures were known to be "clear and comprehensive and his enthusiasm was contagious."[4] Those who worked with him said he was "loved and respected" by his students and "inspired so many able minds who will carry on in the spirit of his work."[5] He continued to research and teach at Columbia until his death.

At Columbia, Wald was often known as "one of the smartest people in the room."[6] One of his students, Allen Wallis, thought so too and arranged for Wald to be drafted into the Statistical Research Group (SRG), a secret, highly classified agency within the Office of Scientific Research and Development (OSRD), which reported to the president.

The SRG was assigned work by the OSRD. Military commanders would contact a member of the OSRD at the White House. A group from the SRG would then travel to Washington to discuss the task, and a determination would be made on whether the mathematicians could help. The SRG's contributions to the war effort were wide ranging and included recommendations on ammunition mixtures, sampling inspection plans for rocket propellant, proximity fuse settings, and other logistical and ballistic challenges.[7]

Wald understood the significance of the tasks assigned to the SRG, and others wrote of the "powerful, pervasive, and unremitting pressure."[8] The work must have weighed heavily on his shoulders.

Milton Friedman, one of the original members of the SRG, in his autobiography more than four decades later wrote of how those in the group were bound together by common cause:

> (We) were all cooperating with one another under intense pressure with a single objective: contributing to the effectiveness of our fighting forces. I saw less internal bickering, less office politics, less self-interested manipulation, more concentrated hard work, during that period than I have in any comparable period before or after.[9]

Friedman understood the importance of many of the discoveries in mathematics, such as the invention of sequential analysis, that occurred at SRG during the war years. He also grasped the purposes for which the discoveries were developed. Later in his life, reflecting on his time at SRG, Friedman said: "Ghastly that such progress should have involved, even depended on, the deaths of millions of people thousands of miles away."[10]

Identifying WWII Aircraft Vulnerabilities

Much has been said about the events described in this section, and much that has been said is not true. I will not try to correct other accounts but will simply lay out what I have discovered based on written accounts by those with firsthand knowledge of the principals and the now declassified materials contained in the US National Archives.

One of the tasks assigned to SRG was to determine the parts of Allied aircraft that were most vulnerable to enemy fire from the ground and air. This was important because the carnage of men and machines over the skies of Europe at that time was enormous, at levels unimaginable for today's military. A bomber crew had a less than 50 percent chance of surviving the standard twenty-five-mission tour of duty.[11] Some individual sorties suffered losses of one in five planes.[12] The Fifteenth Air Wing lost 5,090 of 6,858 aircraft during the war.[13] Even if a plane returned safely, some of the crew often did not.

A historian wrote later of the Allied airmen:

> *It wasn't just the likelihood of death but the method of dying that kept young airmen awake in their billets. The choices were stark. If they didn't perish from searing flak or fighter bullet, they might fall without parachutes from suddenly disintegrating aircraft, or drown in the cold waters of the North Sea . . . the likelihood of survival in the bomber was so slim they were no more substantial than men of air, ghosts already, waiting to vanish.*[14]

The German military devoted vast amounts of resources to defending the Fatherland from Allied aircraft. At the height of the Allied bombing campaign, Germany had two million soldiers and civilians devoted to antiaircraft defense, more people than all those employed in the Nazi aircraft industry.[15] An estimated 30 percent of all guns and 20 percent of all heavy ammunition manufactured in Germany during the war was for antiaircraft weapons.[16] The German military commanders had no choice—Allied bombing was destroying wartime manufacturing and killing civilians. By the end of the war, Allied bombers had dropped two million tons of bombs and leveled more than sixty cities.[17]

The bombing by US and British forces played a significant role in the defeat of Germany. But the men in these aircraft paid a high price.

A nineteen-year-old flight engineer recalled:

> When we first arrived on 101 Sqn the intelligence officer told us, "You're now on an operational squadron, your expectation of life is six weeks. Go back to your huts and make out your wills." It was simply accepted that two out of three of us would be killed.[18]

By the end of the war, the Allies had lost over 33,000 planes and 160,000 airmen.[19]

Wald knew well the death toll of US and British airmen in the skies over Germany. He was also deeply concerned about his parents, brothers, and sisters back in Germany. (He would only learn of their fate later.)

When Wald was assigned the task of identifying the parts of Allied aircraft that were most vulnerable to enemy fire, he began by segmenting the data and counting the number of hits sustained by different parts of the plane. He assumed all undamaged bombers would return and that bombers that sustained more than a certain number of hits in a given section would not. He adjusted for the varying surface areas of each section. The expansive wings of military aircraft, for example, were more likely to be hit than the narrow cockpit. He compensated for different types of guns. German cities shot varying caliber weapons, and Allied aircraft could be strafed by bullets from enemy fighters. Depending on the type of projectile, he varied the vulnerability of a plane to a given number of hits.

After sorting and placing this data in tables, Wald invented a number of statistical techniques to estimate the vulnerability of Allied aircraft. These techniques are beyond the scope of this book. But decades later, mathematicians still marvel at his achievement.

A prominent statistician wrote:

> *Wald's work on this problem is difficult to improve upon . . . By the sheer power of his intuition Wald was led to subtle structural relationships and was able to deal with both structural and inferential questions in a definitive way.*[20]

Despite inventing groundbreaking statistical techniques, Wald is most well-known today for one critical insight concerning where to place protective armor on the bombers. This insight was the basis of a controversial recommendation.

Wald told the military commanders that the most vulnerable parts of the planes were where the bullet holes were not.

The generals initially disagreed. Skepticism on the part of the military about Wald's recommendation was reinforced by the fact that it came from an Austrian professor, with no combat experience, working out of New York City, thousands of miles from the skies of Europe. But Wald was able to convince the military that the parts of the plane with the least damage were the most vulnerable. Wald realized that the planes that did not return had been struck in the areas that showed the least damage on the returning planes. This explained why the returning planes had sections of the aircraft that were largely unscathed.

The story of Wald and the Allied bombers is the most often cited example of survivor bias. In this case, the "winners" were the planes that returned safely to base. The "losers" were those that had been shot down over Germany. Wald was able to infer from the surviving planes what caused the non-survivors to crash. Wald understood he was working with an incomplete data set with no means to retrieve information on the non-surviving planes. But the distribution of the damage on the returning planes allowed Wald to characterize the differences between the surviving and non-surviving aircraft. Wald was not fooled by the winners, in this case the planes that returned safely to base.

Wald did have to make several assumptions. Aircraft could also crash due to mechanical failure, pilot error, running out of gas, or inclement weather. These or other factors were probably the cause of some of the missing planes. However, Wald reasoned that if one or more of these factors was the primary reason Allied aircraft did not return safely to base, then the damage to the surviving planes would have been evenly distributed. Wald was able to complete the sample set by correctly estimating the characteristics of the non-survivors, the planes that sustained fatal damage and crashed.

In the years that have followed WWII, numerous articles claimed that Wald concluded the most vulnerable parts of an aircraft were the engines. In fact, this was not the case. The source of this confusion comes from the two reports Wald wrote on the subject.

The first report was an internal memo written by Wald in 1943 entitled *A Method for Estimating Plane Vulnerability Based on Damage of Survivors.*[21] Released after WWII, this memo was unclassified because it did not contain actual data on damage to Allied aircraft, which the military considered sensitive information. Rather, this internal memo describes only the mathematical theories Wald developed. As part of this memo, Wald did illustrate his theories in one example by creating hypothetical data on damage to Allied aircraft.

The hypothetical data Wald generated divided the surface area of Allied aircraft into four sections:

1. Engines

2. Fuselage

3. Fuel system

4. Other parts

Based on this hypothetical data, with assumptions for a certain number of hits in each of the four sections, Wald showed that the probability of being downed by a single hit to an engine was 39 percent. In contrast, the same probability for the other parts of the aircraft was substantially lower: fuel system (15 percent), fuselage (5 percent), other parts (2 percent). Accounts of Wald's work reference this first report, published online by the Center for Naval Analyses and available to the general public.

However, Wald also wrote a second report two years later, in 1945, entitled *The Estimation of Vulnerability of Aircraft from Damage to Survivors.*[22] This report was classified because it contained actual combat data on damage to Allied aircraft. Now declassified, this second report provides a more nuanced, complex analysis and gives greater insight into Wald's thinking and actual conclusions.

At the beginning of the 1945 report, Wald relates a story that he heard from his days in Austria before immigrating to America and contrasts it with the approach he developed.

Wald writes:

> *An Austrian physician is reputed to have estimated vulnerability of soldiers in World War I from hospital records, and to have concluded that injuries to limbs were most dangerous and injuries to the head least dangerous because so many hospitalized soldiers were wounded in the limbs and so few in the head. Our method of estimating vulnerability is essentially similar—except that we would draw the opposite conclusion . . . if aircraft never return with hits on the oil cooler, but with many on the vertical stabilizer, we infer that the oil cooler is highly vulnerable and the stabilizer is not . . . In practice aircraft return with hits on every component . . . These survivors give us information on vulnerability.*[23]

In the second report, Wald used actual data from classified, highly sensitive intelligence gathered from August 24, 1943, to May 31, 1944, on P-47s and P-51s in combat. The P-47s and P-51s were single-engine fighters from the Eighth Fighter Command. There is no evidence that Wald made recommendations for other planes, such as B-17s or B-29s, the four-engine bombers deployed by the military at that time. This was because Wald found difficulties with adjusting his analysis for duplication of vulnerable parts. B-17s and B-29s could fly with one or more damaged engines.

Wald cautioned that:

> No workable method has yet been found for esti-
> mating the vulnerabilities of duplicated parts. This
> means at the present the method is practicable
> only for single engine fighters, twin-engine aircraft,
> which rarely or never can return on one engine.[24]

The data Wald used is laid out in Table I of his second report.[25] During the period from August 1943 to May 1944, P-47s recorded 104 losses out of a total of 1,912 aircraft. The same numbers for P-51s were 82 and 1,257, respectively. Wald sectioned the areas of the P-47s and P-51s differently than in his first report. More importantly, the distribution of hits on various areas of the plane in actual combat differed from the hypothetical data he supplied in the first report.

The distribution of hits based on combat data for P-47s and P-51s was:

- Wings: 33 percent

- Fuselage: 22 percent

- Cockpit: 7 percent

- Tail section: 19 percent

- Engines: 19 percent

Wald was skeptical of these results and wrote:

> These conclusions are hardly plausible, since other
> evidence suggests the engine is relatively vulnerable

*and the fuselage (even including the fuel system)
is not . . . it is hard to see how the average presented area of the engine to attacks chiefly from
astern can amount to 19 or more percent of the
total presented area . . . the enumeration of hits on
surviving aircraft also appears to be based upon
point of entrance of the projectile, so partially
shielded areas, such as the cockpit, will be scored
with fewer hits than they received.*[26]

In a final section of the second report, Wald also considers the
vulnerability of P-47s and P-51s to 20mm and .303-caliber projectiles. For 20mm projectiles, the data shows the cockpit and fuel
system (which is part of the fuselage) are the most vulnerable, followed by the engines and wings. For .303-caliber projectiles, the
data demonstrates the fuel system is the most vulnerable, followed
by the engine, then cockpit and wings.

Given these somewhat contradictory and counterintuitive
results, it is not surprising the mathematician was reluctant to
draw firm conclusions. Perhaps he began his work with the pre-
conception that the engines were the most vulnerable. This would
explain the hypothetical data he created in the 1943 memo, which
focuses on the importance of protecting the engines. But the first
report was written before Wald had actual data from P-47s and
P-51s in combat.

Wald wrote in both reports that he was convinced that the most
vulnerable parts of an aircraft were those with the least damage
and that survivor bias accounted for this apparent contradiction.
However, nowhere in these two reports does Wald make a specific recommendation on which parts of fighter aircraft are most
vulnerable and should be protected. He also does not address the
issue of the vulnerability of B-17 and B-29 bombers. In the first

report he did not have the data to make a recommendation and, in the second report, the data he had contradicted his intuitions.

Therefore, it is not surprising that Wald in the end was silent on the issue of the most vulnerable part of any Allied aircraft, whether fighters or bombers. And despite subsequent reports to the contrary, additional protective armor was not installed on military aircraft during WWII based on Wald's findings, even on fighter planes. For example, the P-51 had protective armor only behind the seat and headrest of the pilot and in front of the engine. This did not change throughout the production run of the fighter.

Nevertheless, Wald's insight that the most vulnerable part of an aircraft was the section with the least damage became an important part of aircraft design. Based on this insight, the vulnerability of military planes was reduced by avoiding "single-point-kills" in which the plane could be downed by a single projectile. Because of Wald's work, cooling and lubrication systems were moved deeper into the wings and fuselage, self-sealing fuel tanks were installed, and air-cooled engines were used to reduce the number of critical components.[27]

The standard military reference book on aircraft design states: "The lessons to be learned . . . are that single-point-kill design must be eliminated early in the design of an aircraft and that this no single-point-kill design requirement will reap benefits in both peace time and war time far exceeding any costs incurred."[28]

In a sad and ironic final note, Wald and his wife died in a plane crash in India on December 13, 1950. His book on statistical decision functions had just been published, and Wald had been invited to give a series of lectures by the Indian government. But the Air India flight, lost in a fog, crashed into the peak of Nilgiris, killing all aboard.

His former colleague at SRG, Jacob Wolfowitz, wrote of him after his death, "All must mourn for the statistical discoveries yet unmade which were buried in the flaming wreckage on a mountainside in South India."[29]

Wald's work at SRG is the most often cited and most famous example of survivor bias. However, even prior to WWII, other examples of survivor bias in warfare had been identified.

WWI Helmets

During the first years of WWI, British soldiers marched into combat with soft, peaked fabric caps covering their skulls. German soldiers were similarly minimally protected, wearing the Imperial Army spiked *Pickelhaube*, a hard leather bowl with a brass spike on top for ornamentation.

But head injuries due to airborne shrapnel during WWI became a significant cause of death for troops hunkered down in the trenches. So, the warring armies designed and distributed metal helmets to better protect soldiers' skulls. The British metal helmet was known as the "Brodie," the French helmet was known as the "Adrian," and the German helmet was known as the "Stahlem."[30] The Brodie was considered the best of the bunch. The bowl of the Brodie was constructed of a manganese steel alloy with 1 percent carbon for higher impact strength. Some estimate the Brodie offered 50 percent more protection than the Adrian.[31]

The Brodie was first widely deployed in April 1916 at the Battle of St. Eloi. But there were not enough helmets for all the troops on the front lines. It was not until the summer of 1916 that the Brodie became standard issue.[32] The Brodie would continue to be worn by British troops through the 1930s.

Once the Brodie became standard issue, the British Army doctors noticed a significant increase in head injuries.[33] This caused some to question the effectiveness of the new metal helmets, as the net result of equipping troops with better head gear seemed to be more head injuries.

Military doctors eventually realized the increase in head injuries was due to survivor bias. The number of soldiers struck in the head by airborne shrapnel and other projectiles had not changed. But soldiers who previously would have died wearing fabric helmets now survived due to the superior protection offered by the Brodie.

Tourniquets

A more recent example of survivor bias in warfare is the use of tourniquets on the battlefield.

The Iraq and Afghanistan Wars changed the policy of the US military concerning tourniquets.[34] Before these conflicts, tourniquets were recommended only as a measure of last resort in treating war wounds, and their use was discouraged by military doctors because a tourniquet left on too long can result in the loss of a limb. Also, most tourniquets were improvised from whatever materials were around, so their effectiveness was uneven and inconsistent.

In 2005, this policy changed. Due to a significant rise in the number of severe injuries from roadside bombs, or IEDs, and advances in tourniquet design and manufacture, the surgeon general of the US Army recommended that tourniquets be issued to all soldiers. Soon, US Army personnel were strapping these modern tourniquets around their arms and legs before heading out on high-risk missions.

The US Army collected data on survival rates of injured soldiers who arrived at hospitals with and without tourniquets in place. The results surprised army doctors: The survival rates were comparable for both groups, suggesting tourniquets were of no benefit.

At first, the doctors believed this unexpected outcome was because the more severely injured were more likely to wear a tourniquet. They thought that the benefit of tourniquets was masked in the data because those with more severe injuries were more likely to wear tourniquets but also more likely to die. So, the doctors segmented the data by severity of injury and then compared survival rates of those who arrived with and without tourniquets. And again, they found no difference.

The doctors were perplexed. They knew that modern tourniquets were saving lives, but the evidence they had indicated otherwise.

But the doctors had not considered survivor bias.

The doctors had gathered data on injured soldiers who were admitted to the hospital—those who had survived long enough to make it back from the battlefield. However, doctors had not taken into account the non-survivors, those who were severely injured but were unable to put tourniquets in place and expired before reaching the hospital. Once admitted to the hospital, the lifesaving measures undertaken were the same for those with and without tourniquets, and hence survival rates post-admission were comparable. But the doctors were focused on the survivors, compiling data on mortality rates on those who lived long enough to be admitted. Once the non-survivors were included, those who died before reaching the hospital and never admitted, the benefit of tourniquets was clear.

After correcting for survivor bias, the military realized that tourniquets saved lives not by reducing the severity of injuries but by allowing soldiers to live long enough to make it to the hospital.

Conclusions

Abraham Wald pointed out how even the best minds of the US military can be fooled by the winners. His analysis showed that a careful study of survivors can tell us a lot about the non-survivors, who are sometimes difficult or impossible to observe. Most importantly, he demonstrated that we should pay close attention to both survivors and non-survivors before drawing conclusions.

In the case of military aircraft, the non-survivors were the planes that did not return to base. In the example of WWI helmets, the non-survivors were the doughboys who would have died without the added protection of the Brodie. In the instance of tourniquets, the non-survivors were the soldiers who perished in transit to the hospital because they were not wearing a tourniquet.

The impact of survivor bias in warfare is real. Abraham Wald showed us how easily we can be deceived by survivor bias and how to avoid being fooled by the winners.

In the next chapter we will continue the WWII theme, but our focus will be different. During and after WWII, there have been examples of victors erasing the vanquished from our collective memory. Written records are destroyed, testimonials are suppressed, and those who try to speak for the dead are persecuted. Even today, significant events in history have been shoved into the shadows, relegated to dark corners of our past, or distorted to suit the purposes of those who ultimately prevailed.

In these cases, we are fooled by the winners because the winners can rewrite history.

CHAPTER 6

History: Winners Hold the Pen

WINSTON CHURCHILL IS COMMONLY GIVEN credit for coining the phrase, "History is written by the victors."[1]

In fact, there is no evidence he actually spoke or wrote those words. Churchill did say in a speech before the House of Commons on January 23, 1948, "For my part, I consider that it will be found much better by all parties to leave the past to history, especially as I propose to write that history myself."[2] But decades before Churchill, Napoleon probably said it best: When asked about how he would be remembered, Napoleon reportedly replied, "History is a set of lies that people have agreed upon."[3]

Following are examples of some lies that persist to the present day.

Dr. Shirō Ishii and Unit 731

From 1932 to 1945, the Japanese Army undertook the largest biological warfare program in history. This biological warfare program, funded with secret monies from the Japanese Army military budget, was carried out by Unit 731. The creator and leader of Unit 731 was Dr. Shirō Ishii (1892–1959).

Ishii joined the Japanese Army in 1922 and was an early advocate of creating bioweapons. Ishii argued that "Japan did not possess sufficient natural resources of metals and other raw materials required for the manufacture of weapons." Because of this limitation, Japan should "develop new types of weapons . . . in the sphere of bacteriological warfare."[4] Initially Ishii's proposals were ignored. In 1928, there was a change of leadership in the Medical Division of the Japanese Army, and he was sent on a two-year overseas tour to gather intelligence on biological warfare from Western militaries. In 1931, he was promoted to senior army surgeon, and the following year he was sent to Manchukuo, an area that is now part of China, to build a secret biological weapons program.

Ishii's first years in Manchukuo were taken up with constructing factories for manufacturing biological weapons and laboratories for human experimentation. During this time, he also invented a water filtration system that became standard equipment for the Japanese Army throughout the world.[5] In 1933, on a return trip to Japan, he demonstrated his system to Emperor Hirohito at the Army Medical College. Ishii urinated into one of his filters, poured out a cup, and then offered the emperor a sip. The emperor demurred, whereupon Ishii promptly swallowed the filtered urine.[6]

The center of Ishii's operations and the headquarters of Unit 731 was the human experimentation laboratory at Ping Fan, in Northern Manchuria. The Ping Fan complex was spread out over 800 acres and heavily guarded. Local peasants were warned that anyone approaching Ping Fan would be killed on sight.[7] Japanese fighter planes patrolled the skies over Ping Fan from a local air-field and had standing orders to shoot down commercial aircraft that strayed over Ping Fan airspace.[8] At the center of Ping Fan was a facility known as the Zhongma Fortress, a castle-like complex surrounded by a ten-foot-high wall topped with high-voltage barbed wire.

The subjects of the experiments at Ping Fan were prisoners of war (POWs) transferred from Japanese prisoner of war camps throughout Asia, and Chinese civilians kidnapped from local farms and villages. Ishii's victims included American and other Allied POWs. The prisoners generally lived no more than a few weeks after arriving at Ping Fan and, once no longer useful as test subjects, were cremated in one of Ping Fan's large furnaces.[9] The bodies generally burned quickly because the internal organs were gone, removed without anesthesia because Ishii believed painkillers interfered with lab results.[10] Those few who survived experimentation were shot, injected with poison, or killed with a blow to the head with an axe.[11]

The tests conducted by Unit 731 at Ping Fan included infecting prisoners with plague, cholera, typhoid, dysentery, glanders, tetanus, tuberculosis, anthrax, and gangrene. Sometimes the prisoners were given forced injections, and other times food was laced with pathogens. Chocolates and cookies were used for infecting children.[12] Other experiments included stress tests, such as hanging people upside down until they choked to death, or injecting air into blood vessels to induce embolisms.[13] A special pressure chamber was constructed to determine how much pressure a human body could withstand until it collapsed and how little pressure a human body could endure before it burst.[14]

Ishii was interested in how bioweapons affected not only Chinese but also Anglo-Saxons, in anticipation of an invasion of the United States.[15] At least 1,300 American POWs captured in the

Philippines were transferred to Mukden, another of the experimentation camps run by Unit 731.[16] The diary of a US Army captain records that he and his fellow American prisoners at Mukden were given injections of smallpox, typhoid, dysentery, and cholera.[17] His account is corroborated by statements from British POWs and former workers at the Mukden camp.[18]

Ishii's research into bioweapons went far beyond forced injections and dissecting fully conscious human subjects in the labs of Unit 731. Ishii also tested large-scale deployments of bioweapons on Chinese populations. Unit 731 used bioweapons in at least eleven reported attacks on Chinese cities.[19] In 1938 and 1939, planes dropped a jelly-like substance laced with plague and cholera over parts of Yangjin in the province of Guangdong. The resulting cholera epidemic killed over one million people.[20] In 1940, Unit 731 sent planes to disperse insects infected with plague over the skies of Ningbo and trains to dump liquids containing cholera and typhoid bacteria into the water supply, precipitating multiple epidemics.[21] In 1941, Unit 731 dropped wheat and corn infested with plague on the city of Changde in Hunan Province.[22] Areas in China in which Unit 731 bioweapons were deployed continued to experience unusually high levels of infectious diseases for almost a decade after WWII.[23]

As WWII was coming to a close, Ishii planned a desperate strike at the United States mainland. Code named "Cherry Blossoms at Night," Ishii's plan called for five I-400–class submarines carrying Aichi M6A Seiran aircraft to traverse the Pacific Ocean and then surface off the coast of California. Kamikaze pilots would then fly the aircraft through the dark night to drop their deadly loads of plague bacteria into the heart of San Diego. The attack was scheduled for September 22, 1945.[24]

On August 6 and 9, 1945, the United States dropped atomic bombs on the Japanese cities of Hiroshima and Nagasaki. On

August 10, a special messenger arrived at Ping Fan from Tokyo with instructions that Ishii was to "eliminate all evidence of biological warfare from the Earth forever."[25]

Ishii's first task was to have his troops shoot the remaining prisoners still alive and then burn the bodies. This took three days.[26] Next, Ishii destroyed as many buildings as possible, including the human experimentation laboratories. But Ishii saved many of the jars and slides with human specimens as well as crates full of reports documenting the results of his experiments and shipped them back to Tokyo. Before departing, Ishii released the remaining disease-infected fleas and rats from Ping Fan into the countryside, sparking an epidemic that killed 20,000 to 30,000.[27] Ishii then boarded a plane back to Tokyo, and his family and other members of Unit 731 got on a train bound for Korea and were subsequently evacuated by ship to Japan.[28] On August 15, Japan surrendered.

Upon his return to Japan, Ishii was concerned that he would be tried as a war criminal. So, he arranged for friends to plant a story in the newspaper that he had been shot. In November 1945, his friends staged a mock funeral, complete with wailing mourners and solemn priests.[29]

Ishii had stolen large amounts of money from the Japanese Army at the end of the war and used these funds to pay others to keep him hidden.[30] However, numerous stories about Ishii and the atrocities of Unit 731 began finding their way back to Allied command in Tokyo. A US intelligence report dated January 7, 1946, claimed that Ishii had embezzled $1 million in cash from the Japanese Army and stated that he had "committed biological experiments in Manchuria and should be arrested and interrogated."[31] US authorities began searching for the doctor.

On February 5, 1946, Ishii was captured and confined to his home in Tokyo, awaiting further investigation. A military tribunal in Tokyo convened on August 29, 1946. The tribunal heard testimony from one of the American prosecutors who had been charged with investigating Unit 731.

The prosecutor stated:

> *The enemy took our countrymen as prisoners and used them for drug experiments. They would inject various types of toxic bacteria into their bodies and then perform experiments on how they reacted, sacrificing our brothers . . . this is treatment which would not even be given to dogs and cats.*[32]

During 1946, Ishii and other members of Unit 731 were interviewed multiple times as war criminals. However, Ishii denied anything other than some research on animals to test the effectiveness of various vaccines. Then, in March 1947, Ishii surprised his American interrogators by offering them a deal. Ishii said: "If you will give me documentary immunity for myself, superiors, and subordinates, I can get all the information for you. I would like to be hired by the US government as a biological warfare expert."[33]

US intelligence agencies recommended the United States accept Ishii's offer. Part of a memo sent back to Washington admitted that Ishii and his team of doctors had committed war crimes, but nevertheless recommended the members of Unit 731 not be prosecuted:

> *The BW information given by the Japanese as well as the information which it is expected they will give is considered of vital importance to the security of this country by both the Chemical Warfare Service and the Intelligence Division . . . the value to the US of Japanese BW data is of such importance to national security as to far outweigh the value accruing from "war crimes" prosecution.*[34]

This recommendation was seconded by US Army doctors stationed at Camp Detrick, Maryland, responsible for the US biological warfare program. These doctors were concerned the US had fallen behind the Soviet Union in developing biological weapons and viewed Ishii and the materials in his possession as a way to leapfrog ahead.

The prosecutors charged with investigating Japanese war crimes strongly disagreed. The US State Department also vehemently objected, warning that an immunity deal with Ishii "might later be a source of serious embarrassment to the United States."[35]

The memos outlining these disagreements within the US government were sent to General MacArthur and copied to the commander in chief, President Truman. There is no direct evidence Truman read these documents. However, given the strong disagreements between the State Department and the American war crimes prosecutors on the one side, and the Department of War and the intelligence agencies on the other, it seems unlikely Ishii and his men could have escaped the war crimes trials ongoing in Japan without Truman's blessing. Furthermore, Truman was a strong backer of the army's biological warfare program at Camp Detrick, whose doctors desperately wanted to get their hands on Ishii's data and human specimens.

In the end, Ishii and his men were granted immunity and were never tried for war crimes by American forces or the Japanese government. In return, Ishii guided American Army officers to temples where he and his officers had buried thousands of microscope slides containing slivers of kidneys, livers, spleens, and other organs from human experiments at Ping Fan. These slides were shipped to Camp Detrick, Maryland.[36]

Upon receipt of the material, the US Army doctors at Camp Detrick reported back: "Information with respect to human susceptibility to these specific infectious doses of bacteria . . . could not be obtained in our own laboratories because of scruples attached to human experimentation."[37]

US intelligence agencies had initially seized evidence of the genocide and war crimes committed by Unit 731 in Tokyo at the end of the war and shipped them to Washington, DC. But after Ishii and his men were granted immunity, the documents were returned to the government of Japan without making copies.[38] The few American POWs who survived captivity in Unit 731 camps were required to sign statements that disclosing their experiences as prisoners and subjects of human experimentation would subject them to court martial.[39] In the years after the war, MacArthur and US intelligence agencies protected Ishii and his men living in Japan. Large payments were reported to have been made to many Unit 731 members during 1948 and 1949, in some cases up to two million yen, a huge sum in war-ravaged Japan.[40]

It has been alleged that Ishii traveled to the United States during the 1950s to work with the US government on bioweapons.[41] The US Army has denied it has evidence that Ishii visited US biological weapons facilities.[42] US intelligence agencies have not been willing to release materials related to Ishii.[43] It was recently revealed that after the end of WWII and continuing into the 1960s, the CIA established "black sites" staffed by Japanese doctors in several East Asian countries for conducting biological experiments on human subjects.[44] It is not known whether Ishii or other former members of Unit 731 conducted some of these experiments.

Ishii lived out his remaining years in relative luxury in Tokyo with funds he had embezzled from the Japanese Army, a monthly retirement check from the government of Japan as a former general, and allegedly payments from the US government. He surfaced once publicly on August 17, 1958, at a gathering of former Unit 731 members and delivered a speech. He died on October 9, 1959, from throat cancer, aged sixty-seven.

Today, Japanese textbooks typically do not mention Unit 731. The Japanese Ministry of Education, which approves all school textbooks, has said, "No credible scholarly research, articles or book, have yet been published on this issue."[45] Authors who have attempted to include accounts of Unit 731 were told by the Ministry of Education to delete those specific passages.[46]

American textbooks are filled with accounts of the horrors committed by Nazi doctors. But there are no similar accounts of Ishii and Unit 731. A survey showed that most textbooks do not mention Unit 731, despite the fact that the biological experimentation on humans was similar to experiments conducted during the Holocaust. The authors of the survey note that most high school teachers are probably unaware of the history of Unit 731, in stark contrast to their knowledge of the Holocaust.[47] This survey concluded that a significant challenge to reporting on the atrocities of Unit 731, as compared with the Holocaust, is due to "the US government's handling of the two subjects in the post-war period— facilitating exposure of the Holocaust while imposing secrecy and silence on the matter of Unit 731."[48]

We began this chapter with the quote, "History is written by the victors." The story of Unit 731 may be the exception: Victor and vanquished colluded to cover up war crimes and genocide carried out with biological weapons.

Truman and the Decision to Bomb Hiroshima and Nagasaki

Another example from the WWII era of the rewriting of history is President Truman's decision in August 1945 to drop atomic bombs on Japanese cities. As a result of the president's decision, over 200,000 Japanese civilians were slaughtered. This decision was justified at the time as necessary to prevent the deaths of tens of thousands of American soldiers that would ensue from an invasion of the Japanese islands.

President Truman decided to detonate atomic bombs over Hiroshima on August 6, 1945, and Nagasaki on August 9, 1945, without consulting broadly with his advisors.[49] The majority of the members of the Joint Chiefs of Staff did not believe at the time that dropping atomic bombs on Japan was warranted: Of the eight five-star generals in the US military only one—Marshall—thought the bombing was justified.[50] Postwar appraisals of Truman's decision by the leading generals of the US military—Arnold, Eisenhower, Halsey, King, Leahy, MacArthur, and Nimitz—concluded that dropping nuclear bombs on Japan was unnecessary.[51]

There are several reasons for this consensus against dropping atomic bombs on Japan among military leaders. Both American and Japanese officials understood by the summer of 1945 that Japan had lost the war. The US military had total, unopposed control of the skies and had blockaded the main Japanese island. Because of Allied bombing and the blockade, the Japanese government had been rationing food since early 1945, but supplies were expected to run out by November.[52] In fact, the emperor feared that if the war did not end soon, a domestic revolt would overthrow his dynasty.[53]

To the west, the Soviet Union was poised to invade the Japanese homeland.[54] The Japanese particularly feared Russian occupation. Admiral Toyoda said in a postwar statement that "the Russian participation in the war against Japan rather than the atom bombs did more to hasten the surrender."[55]

Recognizing the inevitable, Japanese officials had made numerous overtures to the US government.[56] Throughout 1945, the emperor and his advisors sought assistance from Sweden and Portugal to negotiate a peace treaty with the United States. On May 5, a cable from a member of the Admiral Staff of the Japanese

Navy stated that much of the Japanese military "would not regard with disfavor an American request for capitulation even if the terms were hard."[57] On July 13, another cable from the Japanese foreign minister reported that the emperor was "mindful of the fact that the present war daily brings greater evil and sacrifice upon the people of all belligerent powers," and furthermore that he "desires from his heart that it might be quickly terminated."[58]

After Japan's surrender, Secretary of War Stimson commissioned an interrogation of some seven hundred Japanese military, government, and industrial officials and concluded that "prior to 31 December 1945, and in all probability prior to 1 November 1945, Japan would have surrendered even if the atomic bombs had not been dropped, even if Russia had not entered the war, and even if no invasion had been planned or contemplated."[59] Truman was aware of the Japanese overtures and the content of the diplomatic cables. But Truman did not respond to any of them.[60]

Despite what is written in most American textbooks, there were alternatives to dropping atomic bombs on large population centers that incinerated and fatally poisoned hundreds of thousands of civilian men, women, and children.

In my view, the best option would have been to wait. Stalin had committed to Truman to declare war on Japan on August 15, including announcing an invasion of the Japanese islands. The looming prospect of a Soviet invasion might have prompted an immediate Japanese surrender. And Truman did not have to drop the second bomb soon after the first. It was impractical for Japan to negotiate a surrender with an unresponsive US president in the seventy-five hours between the detonation of the first and second atomic bombs.

Of course, it cannot be known for certain how long Japan may have delayed negotiating a surrender to end the war without the bombing of Hiroshima and Nagasaki. The Japanese military was divided. Some wanted to fight on, but many were willing to accept a peace that maintained the emperor's dynasty. (While Truman demanded unconditional surrender, after the war the United States maintained the Imperial Dynasty, and the emperor's grandson sits on the Japanese throne today.) The Japanese also knew that an invasion would likely cost tens of thousands of American lives and believed this grim prospect gave them negotiating leverage. And Japanese scientists at the time had correctly guessed that the United States had only a few atomic bombs in early August. (A third bomb would have been ready by August 21, and more would have followed in September.)[61]

In any case, an invasion of the Japanese mainland would have occurred at the earliest in late October or early November.[62] Given the collapse of food supplies, the threat of an overthrow of the Imperial Dynasty, and a threatened Soviet invasion and occupation, a delay of even several weeks may have resulted in a Japanese surrender. At a minimum, Truman could have paused for more than seventy-five hours before dropping the second atomic bomb.

The fact is that Truman did not hesitate to order the detonation of all the atomic bombs in the military's possession as soon as they were ready. But it is doubtful the Japanese regime could have survived more than several months without negotiating an end to the war. Hundreds of thousands of lives might have been saved if Truman had delayed the detonation of atomic bombs over Japanese cities until just before a planned invasion of the Japanese islands later in the year. And Truman could have responded to the Japanese entreaties for surrender.

But he did not. There is no evidence Truman considered even a short delay or the possibility of peace.[63]

On August 9, after Nagasaki had been incinerated by a second atomic bomb, Truman received a telegram from a Protestant clergyman, pleading with him to stop the bombing. Truman's racist attitude toward the Japanese was no secret. Truman replied, "When you have to deal with a beast, you have to treat him as a beast."[64]

The version of history recorded in most American history textbooks and commonly repeated today is that the decision to drop bombs on Hiroshima and Nagasaki in August was justified in order to avoid further American casualties. But Truman could have waited, even for just a few weeks, and that might have saved the lives of over 200,000 Japanese people. It might have even delayed the start of the Cold War and the nuclear arms race that followed.

Megamurderers: Hitler, Mao, and Stalin

In terms of absolute number of battle-related deaths, the twentieth century was the bloodiest period so far in human history, at 105 million soldiers and civilians.[65] During that same century, Hitler, Mao, and Stalin killed an estimated additional 20 million, 38 million, and 43 million, respectively, from "genocide, politicide, and mass murder, excluding war dead."[66]

Hitler is universally despised both within and outside of Germany. Since WWII, successive German governments have consistently and harshly condemned his actions. I don't know anybody today who would name their son Adolf. In fact, I don't recall ever meeting an Adolf.

But the governments of China and Russia have rewritten history for their own purposes. As a consequence, Mao and Stalin are today viewed favorably by an overwhelming majority of the citizens of their respective nations.

Leaders of the Chinese Communist Party (CCP) have consistently praised Mao. The official party line is that Mao's "merits are primary and his errors secondary."[67] Deng Xiaoping famously said, "Mao's contributions to the Chinese revolution far outweigh his mistakes."[68] The current Chinese President Xi Jinping has called for a "New Long March" in an effort to evoke parallels between himself and Mao.[69] The current leaders of the CCP have whitewashed the deaths of 38 million Chinese because attacks on Mao, the founder of the CCP, threaten their own legitimacy. This is despite the fact that Mao had no reluctance to exterminate tens of millions of his fellow citizens, including many of his CCP party comrades.

In 1950, Mao wrote to party leaders: "Land reform in a population of over 300 million people is a vicious war . . . this is the most hideous class war between peasants and landowners. It is a battle to the death."[70] He followed up on this threat with the execution of over 800,000 landlords during a three-year period in the 1950s.[71]

In 1958, he said in a speech, "Emperor Shih Huang . . . buried alive 460 scholars only, but we have buried alive 46,000 . . . We are 100 times ahead."[72]

Mao established a system of slave labor camps with more than 10 million prisoners at any given time. While Mao was ruler of China, an estimated 15 million died from overwork, exposure to elements, and malnourishment in these hell holes.[73]

Despite his bloody reign, Mao is generally viewed favorably in China today. In an online poll, 82 percent of those surveyed viewed Mao positively.[74] Another survey showed that 85 percent of Chinese believe his merits outweigh his faults.[75] One study concluded that "the country's rulers do not just suppress history, they recreate it to serve the present."[76]

In the case of Stalin, the twentieth century's leading megamurderer, successive Soviet and Russian governments have portrayed his leadership as "heroic" and one of the "bright aspects of Soviet past." This government-sanctioned version of history means that the "millions who perished in waves of political repression [have been] pushed to the margins of collective consciousness."[77]

After Lenin's death, Stalin reinstituted the confiscation of private property, and the resulting fall in grain production caused an estimated six million peasants to starve to death.[78] During Stalin's Terror of 1937–1938, more than one million citizens were arrested, of whom half were executed.[79]

Throughout his reign, Stalin terrorized his people as a means to maintain absolute power. He did this by assigning quotas for arrests and executions. An example of a directive was: "To N.K.V.D., Frunze. You are charged with the task of exterminating 10,000 enemies of the people. Report results by signal."[80]

But the Soviet police had trouble filling the quotas with suspected plotters against the state, and so they often shot those arrested for minor criminal and civil crimes, even "executing mothers and wives who showed up at the police station asking about their arrested loved ones."[81]

However, Russians seem to have forgotten the 43 million of their fellow citizens that Stalin murdered. In 2019, 70 percent of Russians polled report they "approve" of Stalin's role in Soviet history.[82] Those conducting the survey said that Stalin is perceived as a "symbol of justice."[83] Only 13 percent said they dislike the former Soviet dictator.[84]

Mao and Stalin were each responsible for about twice as many murders as Hitler. Yet, Hitler is universally and regularly vilified (as he should be), and Mao and Stalin are spoken of approvingly in their home countries. This is because the political party that Hitler founded lost WWII and is now gone. But Mao's CCP continues to hold a firm grip on China. Vladimir Putin rules Russia through the United Russia Party, a successor to Stalin's Communist Party of the Soviet Union.

The contrasts in how Hitler, Mao, and Stalin are viewed today show how those governments and political parties that survive can rewrite history.

Slavery in the United States

One of the most widely used history textbooks for high school students in Virginia during the 1950s through the 1970s was the *Cavalier Commonwealth*.[85] This textbook tells the history of the state of Virginia from the arrival of the first Europeans in 1606 to the present day, including passages about the enslavement of African Americans. The *Cavalier Commonwealth* perpetuates a number of myths to justify enslaving African Americans and disregards the violent methods used by European White immigrants to retain political power.

The *Cavalier Commonwealth* touts the "advantages" held by African Americans. According to the *Cavalier Commonwealth*, African Americans resident in Virginia "recognized slavery as a stabilizing influence in a society consisting of two races that had attained quite different stages of development in civilization."[86] The *Cavalier Commonwealth* incredibly recounts tales of how African Americans came to appreciate their life in Virginia, before the Civil War disrupted race relations.

The *Cavalier Commonwealth* idealized a world created by White enslavers to justify the oppression of an entire race. It is shocking to read the language contained in the official state-sanctioned textbook taught to all Virginia high school students:

> *We know that there was some cruelty and suspicion. But we know also that the relationship between the two races was governed by much kindness and mutual trust. Both understood that bondage as they knew it was not totally evil; both realized that enslavement in a civilized world had been better for the Negro than the barbarities he might have suffered in Africa; both were aware that it was questionable whether the Negro was adequately*

*prepared for freedom . . . the two races did not
foresee how soon they would be rudely forced into
proving they could live together peaceably.*[87]

According to this textbook, even after the Civil War, African Americans in Virginia understood the benefits of separating the races in daily life. The textbook asserts that African Americans voluntarily decided not to attend White churches.[88] Statements like this completely ignore the racism and violence that was part of daily life for African Americans. It is not surprising that African Americans in Virginia chose to worship separately.

The perspective of the White enslavers has dominated historical accounts of the period ever since the founding of our nation. Those with the political and economic power shaped the historical narrative to perpetuate the oppression of those of a different skin color. Fortunately, many historians are now giving voice to those who have been silenced in the past.[89] Books documenting the reality of life for enslaved African Americans are now available and provide much-needed counterpoints to the racist attitudes expressed in the official, state-sanctioned textbook used to teach high school students American history fewer than fifty years ago.

Genocide of Native Americans

In addition to describing the "experiences" of African Americans, the *Cavalier Commonwealth* also recounts the history of Virginia's Native Americans. The textbook describes the culture and economy of the Native Americans, uncritically reflecting the views of White Europeans who conquered the Americas with lethal consequences for the original inhabitants.

The *Cavalier Commonwealth* completely fails to mention the decimation of Native Americans by European diseases, known as the Columbian Exchange, which began in 1492 with the arrival of Christopher Columbus. By the time Europeans landed on the shores of Virginia, relatively few Native Americans were still alive. But there is not a single sentence in the *Cavalier Commonwealth* that references the fact that European diseases had already killed over 90 percent of the Native American population.[90]

The Native American populations of the Americas had crossed the land bridge from Asia before 12,000 BC and therefore had no contact with Europeans for almost fourteen thousand years.[91] Because Europeans had been living in cities for hundreds of years, most had developed immunity to the crowd diseases, such as smallpox, measles, and plague. Europe had also been actively trading with India and the Middle East, exposing Europeans to germs from faraway lands. When the Europeans arrived on the Atlantic shores of North America, they immediately infected Native Americans with Old World pathogens such as smallpox, measles, and plague, killing most of the original inhabitants of the New World. In the two centuries following the arrival of Columbus, nine out of ten Native Americans had died from disease before Europeans brandished a sword or cocked a rifle.[92]

Many Spaniards and some Native Americans believed that these diseases were divine retributions from their respective deities. The Spaniards viewed pestilence as consistent with an Old Testament sign of God's displeasure with the Native Americans, a just punishment for worshipping pagan gods.[93] Many Native Americans concluded that their gods were angry and had abandoned them.[94] This belief was reinforced by the death of many of the Native American leaders.

A horrifying account of the deaths caused by European diseases is from the *Annals of the Cakchiquels*:

> Great was the stench of the dead. After our fathers
> and grandfathers succumbed, half of the people
> fled to the fields. The dogs and vultures devoured

the bodies. The mortality was terrible. Your grand-
fathers died, and with them died the son of the
king, and his brothers and kinsmen. We are born
to die.[95]

The Christian colonizers and Native Americans alike believed that the European God was on the side of the disease-free White Europeans. In many instances, the Native Americans simply gave up, surrendering to the armies of the Europeans without a fight. Newborn babies were left to die, and suicide was rampant.[96] A Spaniard wrote of a double suicide of a husband and wife in which the husband "struck her in the head with his tomahawk, that she might go with him to the Other World" and of "a young mother, whose husband had recently died of smallpox, who killed her two children and then hung herself."[97] A Spanish explorer wrote, when first visiting the Aztec capital of Tenochtitlan, that "the streets were so filled with dead and sick people that our men walked over nothing but bodies."[98] Francisco Pizarro famously bragged of defeating the Inca armies with 168 soldiers and sixty-two horses, which he claimed was a testament to the morally superior Europeans who fought with God on their side. But the armies of the Incas had been decimated by European germs long before Pizarro waged his first battle.[99]

After the devastation of the European diseases, large portions of the Native American population became resigned to the depredations of the invaders and began to obey them because the White men seemed to be blessed by their God.[100] These beliefs also created fertile grounds for mass conversions by Christian missionaries, further subjugating the Native American populations to control by European priests.[101]

John Winthrop, the first governor of Massachusetts, believed the European diseases were part of God's clever plan to free up property for the faithful. Winthrop wrote that: "The natives, they are near all dead of the smallpox, so the Lord hathe cleared our title to what we possess."[102] By the time the Puritan settlers decided to move beyond coastal New England, the genocide of the Native American population was nearly compete and the land cleared for Western expansion.[103]

Europeans sometimes intentionally spread their diseases to the inhabitants of the Americas. In North America, White men offered blankets laced with smallpox as gifts to the Native Americans. A famous case is Sir Jeffery Amherst, a British military commander, who in 1763 sent infected blankets to people of the Shawnee and Delaware Tribes. Amherst would later brag of his success in wiping out the Native Americans, but it was just a boast. By the eighteenth century, most of the Native Americans in New England were already dead, and many of those who survived had immunity to the European diseases.

Over the course of the Columbian Exchange, European diseases killed more than 56 million Native Americans.[104] But most of the historical accounts from this period were written by European invaders and omit any discussion of the massive scale of this genocide, which killed almost the entire Indigenous population of two continents.

Even to the current day, the US government refuses to acknowledge the most atrocious genocide in recorded history. In 2009, the US Congress passed a joint resolution titled "An Apology to Native Peoples of the United States."[105] The resolution states that the US government "apologizes on behalf of the people of the United States to all Native Peoples for the many instances of violence, maltreatment, and neglect inflicted on Native Peoples by citizens of the United States."[106]

Nowhere in the resolution is the word genocide.

More recently, some Native American voices are beginning to be heard. In 2017, Oglala Lakota poet Layli Long Soldier, a winner of the National Book Critics Award and the Griffin Poetry Prize, called out the US government in a collection of poetry, *Whereas*, in a direct response to the 2009 joint resolution.[107] In this work, she vividly portrays "the language of occupation" and the failure to acknowledge the mass genocide of Native Americans.[108]

Conclusions

The sample set of history is rarely, if ever, complete. Those who survived may share a dramatically different perspective on events in the past from those who did not.

If Native Americans had sailed to the shores of England and France and within two centuries a large percentage of Europeans were dead from diseases brought over from the Americas, then textbooks today would be overflowing with condemnations of the genocide that followed the arrival of those from the New World. If the 38 million people murdered by Mao had risen up and overthrown the CCP and the 43 million killed by Stalin had over-powered the Soviet government, then Mao and Stalin would likely be perceived in their respective countries in the same way Germans view Hitler. If Ishii and the other members of Unit 731 had been put on trial in Tokyo for war crimes, then the largest biological weapons program in history would be common knowledge, and Ishii would have taken his rightful place in history alongside Nazi doctors like Josef Mengele.

Historical events are constantly generating new groups of win-ners and losers. How the views of the winners and losers differ is not hard to reconstruct. But governments and political parties that successfully vanquish their foes frequently rewrite history. As a result, we are fooled by the winners.

PART II

106 FOOLED by the WINNERS

IN PART I, WE WORKED through examples of how Diagoras-type survivor bias is used to deceive us. We saw how we can be fooled when we fail to consider those who lost—the non-survivors.

In Part II, we will examine instances of Sailor-type survivor bias in which we fool ourselves. In the analogy of the Greek ships, the observer is a sailor who mistakenly reasons he will always safely return to port because he has always done so in the past. But the sailor fails to realize he cannot count the instances in which his ship is lost at sea because he is dead.

When viewing our evolutionary history, we frequently make the same cognitive mistake. If we were to rewind the tape of history and replay it hundreds of times, we might find that our species rarely survives. It could be that we arrived at the present day by taking a series of turns down a specific set of roads, and even one instance of turning left instead of right would have led to our extinction.

Perhaps we were destined to be here. Or perhaps it has been just dumb luck.

As we will see in the case of Sailor-type survivor bias, we can deceive ourselves due to a flawed view of our past. It is easier to be fooled by the winners when you are one of them.

We begin Part II by examining human evolution. We will see that *Homo sapiens* was lucky to have evolved during the last 5 percent of the planet's history and that a space rock conveniently took care of the dinosaurs for us. During our short time on earth, at least in evolutionary terms, we have teetered on the precipice of extinction numerous times. Our distant ancestors must have often feared for their lives and known how perilously close we came to perishing as a species.

But there have also been some false alarms.

CHAPTER 7

Evolution:
The Last Humans

William Whiston and the End of the World

The first scientist known to have predicted the end of all life on earth from a space rock was William Whiston (1667–1752). Whiston was a theologian and mathematician who rose to prominence due to his mentorship by Isaac Newton.[1] In fact, Newton liked him so much that he arranged for Whiston to succeed him in 1702 as the Lucasian Professor of Mathematics at Cambridge, one of the most prominent academic positions in the Western world. (Stephen Hawking held the same post in the twentieth century.) Over time, Whiston's views on religion and other matters became increasingly at odds with the Anglican Church and, in 1710, he was stripped of his position at Cambridge. Despite his differences with the church, he continued to be one of the most prominent academics of his era, charging admission to well-attended lectures and collecting royalties from widely read books.

In Whiston's lectures and books, he mixed science and theology. He promoted the new and controversial theories of Isaac Newton and preached that every single word of the Bible (in English) was the inspired word of God. Whenever a conflict arose between his science and faith, the latter prevailed. This sometimes dismayed his fellow scientists. Whiston believed, for example, that the earth had been struck by a comet to mark the Fall of Adam and Eve, and afterward our planet's orbit shifted from elliptical to concentric.

Whiston was particularly obsessed with comets and the role they played in Scripture. His book, *A New Theory of the Earth*, maintained that a comet was the cause of Noah's flood, which he claimed began with a rainstorm on the eve of November 28, 2349 BC. Just as a comet had demonstrated God's displeasure during the time of Noah, Whiston believed so too would a comet usher in the Second Coming of Jesus Christ, followed by the end of all life on earth. He was convinced, however, that God would not surprise the faithful by sending a comet unannounced. Whiston prophesied God would provide signs to true believers, in order to prepare the worthy for a final ascent into heaven.

In 1736, Whiston foretold that a great comet would appear in the skies on October 16 to punish the wicked, signaling the return of Jesus Christ. He said this time God had chosen to cook rather than drown the unfaithful; the comet would strike the earth, engulfing the planet in a global firestorm.

Picking a specific date on which the world will end requires conviction. Such a forecast could subject a person to public ridicule if the predicted apocalypse does not occur. Whiston's reputation was particularly vulnerable given the earlier Rabbit Woman affair in which he also foretold the end of all life on earth.

In 1726, Mary Toft of Godalming announced she had given

birth to a number of rabbits.[2] This report was initially greeted with some skepticism, and so a surgeon to the royal family was sent to debunk her claim. Under the surgeon's careful watch, Mary successfully brought into the world several more rabbits, and the royal doctor verified the bunny births. Upon hearing this surprising and definitely shocking news, Whiston proclaimed that women giving birth to animals was a clear sign that the end of the world was near. The result was a panic throughout the countryside and a boycott of rabbit meat.

At this stage, King George I became personally involved and sent two other royal doctors to witness the miraculous births that were wreaking havoc in His Majesty's dominions. Upon further investigation, it turned out some baby bunnies had been collected from nearby farms and then Mary had inserted them into her vagina. Mary was arrested and convicted of fraud. Whiston's public support of Mary's claims and prediction of the end of the world drew widespread ridicule.

So, Whiston was risking what little reputation he had left with his forecast that the world would end on the evening of October 16, 1736. But with his scientific knowledge, Whiston had been able to accurately predict the appearance of a number of known comets. October 16 arrived and, just as Whiston foretold, a comet lit up the sky. Emboldened with his regained and newfound credibility after the Rabbit Woman affair, Whiston fanned the flames of fear with dire predictions of a coming Great Apocalypse. Soon, a panic gripped London. Political and church authorities issued public calls for calm.

But these pleas were to no avail. A run on the banks followed. Some fled London for the countryside. Others took to boats and barges on the Thames to escape the firestorm predicted to shortly engulf London.

After the comet passed and the panic subsided, Whiston's

reputation was permanently damaged. To make matters worse, in the aftermath of the panic, a popular local writer, Jonathan Swift of *A Modest Proposal* fame, penned what became a best-selling pamphlet that satirized Whiston, titled *A True and Faithful Narrative of What Passed in London, During the General Consternation of All Ranks and Degrees of Mankind; on Tuesday, Wednesday, Thursday, and Friday Last.*

Swift quotes Whiston proclaiming on the eve of world destruction:

> *Friends and Fellow Citizens, all speculative science is at an end; the Period of all things is at Hand; on Friday this World will be no more. Put not your confidence in me Brethren, for tomorrow morning at five minutes after five the Truth will be Evident; in that instant the Comet shall appear . . . prepare your Wives, your Families and Friends, for the Universal Change.*[3]

And then Swift twisted the knife by adding his own commentary:

> *At this solemn and dreadful prediction, the whole society appeared in the utmost astonishment; but it would be unjust not to remember that Mr. Whiston himself was in so calm a temper, as to return a schilling a piece to the youths who had been disappointed of their lecture, which I thought from a Man of his Integrity a convincing proof of his own Faith in the Prediction.*[4]

As far as I am aware, no well-known scientist since has picked a specific date and time of day on which the world will end. Warnings from astrophysicists about comet strikes have been wisely couched in probabilistic terms, laid out over flexible timetables.

The Cretaceous Comet

We are fortunate that a small comet simply passed by us in 1736. We are equally lucky that a large one struck the earth 65 million years ago.

During the Cretaceous Period, a comet six miles wide, taller than Mount Everest and traveling at a hundred times the speed of a jet plane, targeted the earth.[5] The comet (or asteroid; exactly which is still debated) entered the atmosphere, created a vacuum that sucked up hundreds of millions of tons of dirt into the clouds, and then smashed into the Yucatan Peninsula with the energy equivalent of a billion atom bombs, leaving a hole in the ground twenty-five miles deep and more than one hundred miles wide. The resulting fires circled most of the planet, burning forests and plant life, instantly roasting any animal in its path into charred ash. Within minutes, the "ejecta"—liquefied rocks sent into the atmosphere from what is now Mexico—rained down upon the planet, and a hail of lethal projectiles shredded animals not previously incinerated into little pieces of bone and flesh. The shock to the earth's crust set off magnitude 12 earthquakes around the globe that shallowed up parts of whole continents, ignited massive volcanic eruptions that covered millions of square miles of land in molten lava, and triggered tsunamis that flushed sea life hundreds of miles inland. Ninety percent of the biomass of the earth was suddenly gone.[6]

And then it got cold.

The ejecta and dirt in the clouds, smoke from planetary-scale forest fires, and ash from volcanic eruptions blocked the sun's rays. Global temperatures may have plummeted as much as fifty degrees over land and thirty-six degrees over the oceans.[7] Less than 1 percent of sunlight reached the planet surface, halting photosynthesis.[8] Many creatures still alive soon starved to death on frozen landscapes.[9] The earth had gone from fireball to snowball.

Bad News for Dinosaurs: Good News for Us

The dominant land animals on earth before the arrival of the comet were dinosaurs.[10] Right up until that day, the future of the dinosaurs looked pretty bright. They were at the top of the food chain, without equal, and there were no signs of that changing. Nothing indicated that the planet was anything other than a habitat particularly well suited to a dinosaur's way of life.

If *Homo sapiens* had been alive at the time the comet struck, we would not have made it.[11] But some animals did survive. A few tiny dinosaurs with wings were not killed. Maybe they were nesting in caves on that fateful day, or flight allowed them to scour the earth for food, such as seeds or the roasted animal carcasses that littered the earth. In any case, these small, winged dinosaurs managed to find food and evolved to become today's birds. Many species of fish, protected by the vast, deep oceans, also survived. And a few small mammals, about the size of rodents, somehow lived through this global holocaust. The descendants of these rat-like creatures would evolve to write books about dinosaurs and survivor bias.[12]

The king of the dinosaurs was *Tyrannosaurus rex*, or *T. rex* for short. Some forty feet long, weighing seven or eight tons, *T. rex* was one of the largest meat-eating animals to ever walk the face of the earth.[13] Its skull was five feet in length with eyes the size of grapefruits and a jawbone lined with fifty or so knife-sharp teeth.[14] Unlike us, *T. rex* could regrow broken teeth, which was helpful given its table manners.[15] The king of the dinosaurs benefited from exceptional senses: binocular vision thirteen times better than modern humans, and the ability to hear and smell other animals at great distances.[16] Above all, *T. rex* was thought to have been among the smartest animals around at that time.[17]

When a large comet struck the earth, this was bad news for dinosaurs but good news for us. Most believe that *Homo sapiens* could not have coexisted with such carnivorous beasts. We probably would have been a tasty appetizer: a tall, moderately sized, fleshy mammal who ran upright (easily spotted) and could not run that fast (easily caught). We can be thankful our ancestors were small, ground-hugging, sub-snack-sized rodents, not worth pursuing. Probably our ancestors' main concern was to not get accidentally stepped on.

We have not discovered any preserved *T. rex* or other dinosaur brains, so the best evidence we have of intelligence is the encephalization quotient (EQ), the ratio of brain mass to body size. As an indication of intelligence, EQ has drawbacks because the size of the individual components of the brain is important. For example, the size of the neocortex matters more than that of the limbic system as an indicator of intelligence. Nevertheless, EQ has been shown to be roughly correlated with IQ.[18] *T. rex* had an EQ of 2.0 to 2.4.[19] That is about twice that of a dog, more than a chimpanzee, and about one-third of a modern adult human.[20]

If *T. rex* had survived, we can only speculate on how far another 65 million years of brain evolution would have taken their species. When humans first evolved, our intelligence was about the same as that of *T. rex*. During six million years of human evolution, our brain size has almost tripled: Our EQ started at 2.5 and eventually rose to 5.8.[21] But *T. rex* would have had a 65-million-year head start on brain evolution. Even if *Homo sapiens* somehow could have found a way to coexist with *T. rex*, it is not clear which would have become the more intelligent and dominant species.

A Planet Better at Preserving Fossils than Life

Sixty-five million years ago was not the first time most living organisms on earth were exterminated. In fact, the space rock that killed the large dinosaurs was the most recent of five mass extinctions, defined as events in which the majority of species on earth perished.

The "big five" mass extinctions and the percentage of species lost were:[22]

- Ordovician: 444 million years ago (mya), 86 percent

- Devonian: 375 mya, 75 percent

- Permian: 251 mya, 96 percent

- Triassic: 200 mya, 80 percent

- Cretaceous: 65 mya, 75 percent

If we had been around during any of those five mass extinction events, we could not have survived. We are fortunate not to have been alive during the first 95 percent of the earth's 4.5-billion-year history.

In particular, the Permian mass extinction, what has been known as the Great Dying, was a close call for all life on earth. A dramatic global rise in temperatures, due to volcanic eruptions dumping massive amounts of CO_2 into the skies, baked to death almost every land species. Oceans turned acidic, burning through the gills and shells of most sea life. If all life had ended, we don't know how long it would have taken for it to restart, if ever. Even if it had, we don't know how much time would have passed before life once again reached the Permian stage of evolution. The evolutionary climb to the heights of Permian life took over four billion years after the earth formed. Hence, life might not have restarted, or evolution could have followed a different path. It is highly unlikely that any other path would have taken the same millions of twists and turns that yielded *Homo sapiens*. We are lucky descendants of the fortunate small percentage of all species that have survived these mass extinctions.

Homo sapiens has existed as a distinct species for more than 200,000 years. Despite that, modern-day humans, as demonstrated by DNA evidence, evolved from a group of common ancestors who lived just seventy thousand years ago.[23] After numbering perhaps hundreds of thousands of individuals at one time, the number of *Homo sapiens* fell to as low as several thousand individuals during this period.[24] Such a substantial fall in numbers risked the extinction of our species: A population of merely thousands provides insufficient genetic variation for natural selection to adapt to environmental change and can lead to life-threatening genetic deformities from inbreeding.

The causes of this "bottleneck" in the *Homo sapiens* population are still debated. The most likely explanation is that during this time a volcano, whose caldera today is Lake Toba in Sumatra, Indonesia, erupted in the largest explosion on earth in the past two million years.[25] The Toba eruption spit ash into the skies that blocked out the sun's rays, cooling the planet for more than a thousand years.[26] However, some have argued the impact on the climate of equatorial Africa where most *Homo sapiens* lived was not significant.[27] Other reasons for the population bottleneck could be that *Homo sapiens* were outnumbered by other humans, such as the Neanderthals, and may have suffered devastating attacks from these competing human species. Diseases could have also played a role, but determining the presence of pathogens from ancient skeletal remains is challenging. Or maybe it was some combination of all the above. Regardless, *Homo sapiens* barely survived, passing through a tight population bottleneck that could have easily led to our extinction.

As survivors, we tend to believe that the survival of our species was likely, even inevitable. This is natural. We survived the journey, so we are inclined to conclude that the journey must not have been perilous. But that fails to account for survivor bias. The path we took might have been one of the few that did not literally dead end.

What we do know is that most species don't make it. Ninety-nine percent of all species that once lived on the earth are gone.[28] The vast majority of species on our planet have lasted from one to ten million years.[29] Mammals like us are particularly vulnerable, surviving only about a million years on average.[30] Compared with many other species, humans have a larger mass, which requires more energy when we hunt for food—and offers more food when hunted. Our warm blood demands reliable sources of nutrients to maintain a constant internal body temperature. Our reproductive cycle has a high infant mortality rate (and high maternal mortality rate), and we birth helpless infants completely dependent on adults for survival. These vulnerabilities are exacerbated by an extended adolescence and therefore a greater risk of death from predators. In addition, lower birth rates constrain our ability to adapt to a changing world through natural selection. *Homo sapiens* have the potential to procreate about once a year, but cockroaches give birth monthly. Some bacteria divide into new cells every twenty minutes.

The primary reason species don't last is because the earth's physical environment constantly changes. For all but a few species, the speed of that change eventually outruns the pace of adaptation. Charles Darwin called this series of adaptations "evolution," which he defined as "descent with modification." But the earth's physical environment changes faster than just about any organism can adapt.[31] Sometimes the change in the environment is gradual, such as an ice age, and other times sudden and violent, like a comet strike, a massive earthquake, or a volcanic eruption. In any case, the earth is better at preserving fossils than life.

Despite all this, *Homo sapiens* has survived and flourished. We dominate the land masses of the earth from soaring mountains freezing in the clouds to tropical islands baking in the sun. We are the most intelligent species that has ever lived, as complex, symbolic language gives us the amazing ability to share knowledge with billions of our fellow humans and pass on what we have learned to subsequent generations. In the next hundred years, we may even extend that domination to the planets whirling around our sun.

But our journey to today's world was actually quite perilous. Some fifty thousand years ago at least five human species—*Homo denisova, Homo floresiensis, Homo naledi, Homo neanderthalensis,* and *Homo sapiens*—coexisted on earth.[32] Our species is the only survivor, the last humans.

Homo sapiens and Neanderthals: Two Enter, One Leaves

Homo sapiens first appeared in northern Africa around 200,000 BC, the descendants of an early human species, *Homo erectus.* Sometime around 120,000 BC, *Homo sapiens* started trickling up into Europe, and then a large wave migrated north about 50,000 BC. There, we encountered *Homo neanderthalensis,* more commonly known as the Neanderthals.

Neanderthals were by far the most dominant human species at that time, outnumbering all others combined.[33] Neanderthals had left Africa much earlier, about 800,000 BC, and moved up into Europe. By 400,000 BC, Neanderthals were spread throughout Europe and Asia.

Named after the first specimen found in 1856 by miners in the Neander Valley of Germany, Neanderthals shared many anatomical and cultural characteristics of *Homo sapiens.* Neanderthals were comparable in height, although much heavier in the body, with stronger and thicker arms and legs, a protruding jaw, and a thick-walled skull that rode on top of broad shoulders. *Homo sapiens* and Neanderthals were equally advanced. Neanderthal brains were as large as those of *Homo sapiens*,[34] and like *Homo sapiens*, Neanderthals had language skills.[35] Both species shaped stone tools, crafted jewelry, painted cave walls, buried their dead, and lived on comparable diets.[36] Although physically imposing, Neanderthals in other respects were not that different from modern *Homo sapiens*. It has been said that a Neanderthal on a New York City subway would go unnoticed "provided that he were bathed, shaved, and dressed in modern clothing."[37]

Within thousands of years of *Homo sapiens* arriving in Europe, Neanderthals died out, followed by the extinction of the other human species around the globe.[38] The cause of such a rapid extinction of all humans on earth except for *Homo sapiens* is not known for certain. However, archaeologists have put forward a number of explanations. The most commonly accepted theory is that we slaughtered them, or what has become quaintly known as the Replacement Theory.[39]

As *Homo sapiens* moved into Neanderthal territory, the frequency of violent interactions likely increased. There are indications that *Homo sapiens* had better language abilities than other human species, although our language skills at that time would still have been primitive and limited.[40] Consistent with an ability to better coordinate and communicate, *Homo sapiens* might have been more cunning warriors. It appears many Neanderthal men met an unnatural demise. This theory is supported by Neanderthal male skeletal remains that evidence skulls pierced by arrowheads or foreheads caved in by rocks. One survey of male Neanderthal skeletons showed that 40 percent suffered traumatic head injuries.[41] There is also evidence of widespread cannibalism. Neanderthal skulls were often broken in places that would facilitate extraction of juicy brain matter, and long bones were shattered in ways to ease the scooping of moist bone marrow.[42]

Homo sapiens also bred with Neanderthals. DNA evidence confirms that *Homo sapiens* carry genomes from Neanderthals that range from 1.5 percent in Europeans to 2.1 percent in Asians.[43] *Homo sapiens* men may have killed Neanderthal men to gain access to Neanderthal women.[44] Given the small bands of hunter-gatherers in which humans traveled at that time, interbreeding within a group of forty or fifty individuals could produce genetic anomalies. *Homo sapiens* men who reproduced with Neanderthal women gave their offspring an evolutionary advantage.

An alternative explanation is that *Homo sapiens* were better hunters, and eventually the Neanderthals starved to death. It is not clear why that would be the case: Neanderthals were physically stronger and had established themselves in Europe and Asia long before us. Another theory is we drove Neanderthals to extinction because we domesticated wolves (today's dogs) for hunting and the Neanderthals didn't.[45] It is difficult to know just how important dogs were to food gathering or keeping watch for predators, but it seems unlikely that pets—even working pets—are mainly responsible for the extinction of the Neanderthals. Yet another theory is that *Homo sapiens* infected Neanderthals with lethal diseases we carried within us from Africa. Because the Neanderthals left Africa hundreds of thousands of years before us, they may have lost immunity to African pathogens.[46] This would be consistent with other mass human genocides, such as when Europeans arrived in the New World. However, particular diseases are difficult to detect in fossils, and so there is no direct evidence of the "disease out of Africa" theory.

Based on what is known today, the extinction of the Neanderthals was most likely due to *Homo sapiens* slaughtering Neanderthal men and capturing Neanderthal women.[47] With the Neanderthals out of the way, we could have readily vanquished the remaining less-numerous human species employing a similar strategy. However, the Neanderthals could have just as easily won this interspecies battle for survival. In fact, given their vastly greater numbers and far superior physical strength, Neanderthals probably were the odds-on favorite. And then Neanderthals would have been the ones writing books about those primitive, now-extinct *Homo sapiens*.

Furthermore, we did not just kill off other humans. After vanquishing our nearest genetic kin, we proceeded to exterminate many of our animal predators and hunted to extinction those whose meat we fancied.[48]

Originally, when *Homo sapiens* spread throughout the globe, large animals, known as megafauna, roamed every continent. There is not enough space in this book to list the animals we hunted to extinction, but some examples are straight-tusked elephants, woolly mammoths, woolly rhinoceroses, giant deer, cave bears, and cave lions in Eurasia; marsupial lions, giant kangaroos, giant python, two species of crocodiles, and all large flightless birds in Australia; mammoths, giant elephants, giant sloths, giant armadillos, stag moose, mastodons, and the American lion, which was larger than its African cousin in the New World. Before *Homo sapiens* arrived, camels and zebras roamed the plains of North America. Horses thrived for hundreds of thousands of years before we drove them to extinction on the North American continent around 12,000 BC. (The horses of today were later reintroduced into the Americas by Spanish settlers in the fifteenth century.)

By 10,000 BC, *Homo sapiens* had killed off 80 percent of big animal species in the Americas, basically those large enough to be worth the meat.[49] Some animals we didn't hunt to extinction but dramatically reduced their number, and then we put them in game parks or zoos. Lions have declined from 450,000 to fewer than twenty thousand today, and there were once over one million chimpanzees in Africa compared with an estimated 200,000 currently.[50] The total mass of wild animals is now one-sixth of pre-human levels.[51]

We accomplished the extinction of other animal species through a combination of hunting and turning half of the earth's dry surface into farms.[52] Of course, domesticated animals are a different story. Today, just a handful of species—cattle, pigs, sheep, goats, and *Homo sapiens*—make up 97 percent of the land animal biomass on earth.[53] *Homo sapiens* continues to exterminate other species, with the exception of those destined for our table. The current extinction rate for all species is now more than 100 times higher than in the past.[54] The global population of vertebrates has declined by 52 percent from 1970 to 2010.[55] One-quarter of all mammal and 10 percent of all plant species today are threatened with extinction.[56] All these species, including the other humans, have vanished into the mists of history. The 99 percent of all species that once lived and are now dead are not around to contemplate the prospects for their survival.

At one time the future must have seemed bright for many of these now-extinct species. If given to contemplating their fate, the dinosaurs were probably looking forward to dominating the planet for many millions of years to come. Like us, they might have expected to survive because they had always done so in the past. But if a poll were taken of all species, the survivors and non-survivors, on the question of whether the earth is a hospitable place for life, about 99 percent would vote no. The only ones likely to vote yes—the 1 percent—are those that happened to survive.

Conclusions

Our journey as a species to reach the present day has been fraught with many dangers. We have come close to extinction numerous times, like most species borne on this planet. If *Homo sapiens* had evolved before the last 65 million years of the earth's

4.5-billion-year history, we would not have survived any of the mass extinctions. Since then, threats to our survival have come in the form of periodic ice ages and global warmings, as our planet has tried to alternately freeze or boil us to death. About seventy thousand years ago we barely squeezed through a population bottleneck. Some fifty thousand years ago we battled with the dominant human species on the planet and somehow lived to tell the tale.

Therefore, we should be cautious about becoming overly optimistic about our future. If the past is prologue, then the odds of future survival may not be as favorable as they seem. The journey ahead may present just as many, or even more, opportunities to take a wrong turn, leading to joining the other 99 percent of species who are now extinct.

But that is not the perspective most of us have about our evolutionary history. We somehow survived all these threats to our survival in the past and therefore believe we will somehow continue to do so in the future.[57] But this is just an example of survivor bias. Our perspective on our evolutionary history is distorted because we are one of the few species that made it. The species that didn't would have a different view. We can be easily fooled when we are one of the few winners.

Even though our evolutionary journey to the present day was perilous, many argue that the road ahead presents significantly fewer dangers. After all, Homo sapiens currently rule the planet. We have wiped out or put in zoos any species that threatened us. We have developed powerful technologies to bend the physical world to our will and enjoy a standard of living hundreds of times greater than our hunter-gatherer ancestors.

I will argue in the coming chapters that the risks of extinction for Homo sapiens are greater than ever. We will discuss two in particular: nuclear war and global warming. About seventy-five

years ago, we placed in the hands of a few the ability to exterminate billions with atom bombs. Around the same time, we started
to spew large amounts of CO_2 into the atmosphere, which could
someday trigger a runaway greenhouse effect that fries all life on
earth to a crisp. I will argue these new existential threats put us
at greater risk of extinction than at any time since we battled for
survival with the Neanderthals fifty thousand years ago.

Remarkably, the dangers of these new existential threats were
predicted sixty-five years ago by a Hungarian immigrant to the
United States. The man who saw this clearly during the 1950s,
both the upsides and downsides of technology, worked at the
Institute for Advanced Studies at Princeton and was a friend of
Abraham Wald. This professor cautioned that new technologies
have unforeseen consequences.

In 1955, as he lay dying of cancer near the end of his life, this
professor wrote an essay in *Fortune* that laid out his concerns. He
foresaw much of what would happen in the years to come, including the proliferation of nuclear weapons and global warming.

His essay was entitled "Can We Survive Technology?"

CHAPTER 8

Nuclear War: Dumb Luck

John von Neumann and the Intrinsic Danger of Technology

John von Neumann (1903–1957) was considered by many of his colleagues to be the smartest man alive.[1] Hans Bethe, a Nobel Prize–winning physicist who worked with Neumann on the Manhattan Project, wondered whether "a brain like von Neumann's does not indicate a species superior to man."[2]

Despite an early and untimely death from cancer at fifty-four, Neumann made fundamental breakthroughs in mathematics, physics, chemistry, economics, computing, cybernetics, and evolutionary biology. Neumann published over 150 papers in scientific journals during his lifetime. In addition to his theoretical work, he made significant contributions in the applied sciences. He predicted how cells replicate before Watson and Crick discovered

RNA and DNA.[3] He was one of the most important contributors to the Manhattan Project.[4]

Neumann held significant government positions and influenced postwar military and energy policies. He was an advisor to Presidents Truman and Eisenhower and cabinet secretaries of both administrations. Neumann worked for the RAND corporation on US military strategy, including helping to formulate the doctrine of mutual assured destruction (MAD). (In his last days, Neumann was guarded by military security at the hospital, lest he reveal state secrets while heavily medicated.) He was a leading member of the first Atomic Energy Commission, warned of the dangers of global warming due to the burning of fossil fuels, and forecasted the development of artificial intelligence.

Eugene Wigner, a Nobel Prize–winning physicist, who was friends with Neumann and Einstein, compared the two:

> I have known a great many intelligent people in my life . . . but none of them had a mind as quick and acute as Neumann. I have often remarked this in the presence of those men (who knew both) and no one ever disputed.[5]

Besides his extraordinary intelligence, Neumann was well liked by most people with whom he had contact. This accounted for his success working with teams of scientists as well as national political leaders. Laura Fermi, the wife of Enrico Fermi, said of Neumann:

> Dr. von Neumann is one of the very few men about whom I have not heard a single critical remark. It is astonishing that so much equanimity and so much intelligence could be concentrated in a man.[6]

Neumann had some particular habits that puzzled and occasionally endangered those around him. He seemed to not need sleep, or at most only a few hours a night. He preferred to do his thinking bombarded by loud noises, blasting German marching music or turning up the volume of the TV, while solving lengthy, complex equations. To sort out a particularly thorny problem, he once undertook a ten-hour train trip and upon arrival at his destination, hopped another train back. And he was a notoriously bad driver. He would sometimes careen down the middle of the road while reading a book. At Princeton, he was arrested numerous times for reckless driving and ran into trees at high speeds. An intersection near his house in Princeton was nicknamed "Von Neumann Corner" for all the auto accidents that occurred there. He was in traffic accidents in New York City and California. But the importance of his work for the US military meant that he was never jailed, although he was frequently arrested and fined.

In 1955, Neumann wrote an essay for *Fortune* magazine entitled, "Can We Survive Technology?"[7] In this essay, Neumann argued that three forces would come together to threaten the future survival of *Homo sapiens*. First, technology would exponentially increase the amount of energy at our disposal, including the destructive power of nuclear weapons and the burning of fossil fuels. Second, technology would compress the time in which a decision about the fate of billions would have to be made. Third, geographical separation between nations would no longer protect civilian populations from massive casualties because projectiles could traverse the globe.

More than six decades ago, Neumann predicted many of the challenges of technology that confront us today:

> *Greater automation, improved communications, partial or total climate control, though all are*

intrinsically useful, they can lend themselves to destruction . . . Technological power, technological efficiency as such, is an ambivalent achievement. Its danger is intrinsic.[8]

In particular, Neumann was concerned the pursuit of new forms of energy would lead to armaments of vastly greater power whose use we could not control. His fears were based on a long history of our species trying to kill one other with ever more lethal weapons.

Weapons Have Become More Powerful

As hunter-gatherers, humans wielded "shock" weapons, such as clubs, axes, and swords, that were accurate but had a short range and a force limited to the energy that could be generated by muscle. Even during the Middle Ages, the weapons of war remained largely shock weapons plus some primitive "fire" devices, such as arrows and muskets. But these fire weapons had less force than shock weapons and, in any case, were significantly less accurate. Starting about 12,000 BC, with the Agricultural Revolution, the economic surpluses of large agrarian communities allowed nations to field large armies—thousands and sometimes even hundreds of thousands of soldiers, multiplying the scale of warfare. But the lethal force of an individual soldier remained about the same.

All this changed with the Industrial Revolution, which industrialized not only the economy but also the weapons of war. The power of machines exponentially increased the ability of the armed forces to kill and break things. For example, during the American Revolutionary War, approximately forty thousand British and Americans were killed over the course of twelve years. During the American Civil War, from 1861 to 1865, more than 800,000 perished.

One of the many inventions of the Industrial Revolution was the Gatling gun. Designed by Richard Gatling and patented in 1862, the Gatling gun was a multibarreled weapon that fired a single shot, ejected the spent cartridge, loaded a new bullet, and then rotated to the next barrel. This was the first firearm able to sustain bursts of projectiles. Before the Gatling gun, soldiers had to manually reload each round into a musket or rifle.

Gatling said he invented his gun to reduce the number of war casualties. He hoped that his invention "could by its rapidity of fire, enable one man to do as much battle duty as a hundred, that it would, to a large extent supersede the necessity of large armies, and consequently, exposure to battle and disease [would] be greatly diminished."[9]

Of course, his invention achieved the opposite. A nineteenth-century Gatling gun fired 350–400 rounds per minute, whereas a well-trained Civil War soldier with a Springfield rifle could only load and shoot three rounds during the same period.[10] By the time of the Vietnam War, a version of the Gatling gun known as Puff, the Magic Dragon, named for the immense amounts of smoke and fire emitted from the barrels, fired at a rate of six thousand rounds per minute. Other inventions of the Industrial Revolution, such as high explosives and airplanes, have had similar killing power, but at least had other benefits. The Gatling gun is unusual in that it served no other purpose but to maim and kill, despite the inventor's professed good intentions.

The more powerful weapons of the Industrial Revolution not only fired more frequently, but they also launched projectiles with greater lethal force. A bullet from a Civil War musket had the kinetic energy of one thousand joules, a shell from a WWI artillery gun one million joules, and a shell from a WWII heavy AA gun six million joules.[11] A single WWII hand grenade contained TNT with the explosive power of two million joules.[12] One fire-bombing raid over Tokyo in 1945 released bombs with the equivalent of 60 trillion joules.[13]

As a result, the industrialization of war has steadily increased the death count from armed conflict. At the battle of the Somme, Britain lost twenty thousand men in a single day, and millions died in the trenches of WWI, slaughtered by mechanized weapons.[14] During the battle of Verdun in WWI, approximately two hundred artillery rounds were fired for every casualty.[15]

In tandem with the increased lethality of weapons, the number of people killed in wars has grown exponentially. In the Middle Ages, fewer than one million people per century were estimated to have died from wars.[16] Estimated war deaths during the eighteenth century were seven million.[17] That number rose to 19 million during the nineteenth century and then jumped to more than 100 million during the twentieth century.[18] Military and civilian battle-related deaths from WWII alone totaled more than 50 million.[19] Of course, armies had more targets as the population grew over the centuries. But the Industrial Revolution increased not only the absolute amount but also the incidence of mortality, as the risk of dying from war rose on a per-person basis. An individual living in the twentieth century was thirteen times more likely to die from war than someone alive during the sixteenth century.[20]

Wars have also become more injurious during the twentieth and twenty-first centuries. But this is obscured in the statistics that measure those killed in battle. In fact, comparing battle-death across centuries can be misleading.

During WWI, there were nine million battle deaths over four years. During WWII, those numbers were 16 million over six years.[21] But the numbers of those killed substantially underestimates the increasing number and severity of injuries caused by war. Each successive generation of soldiers has been able to survive a greater percentage of what previously would have been fatal wounds because the injured had access to medical care previous generations of soldiers lacked. Of particular importance is the "golden hour," that first hour after injury. Modern transportation, such as helicopters and ambulances, and fully staffed field hospitals enable surgeons to operate soon after a soldier is wounded.

Many soldiers in the past also succumbed to illness, a rarity on modern battlefields with proper hygiene and effective medicines. In the Spanish-American War, during a six-week campaign in 1898, there were 293 casualties from battle and 3,681 from illnesses.[22] Modern-day soldiers are also protected by more and better armor: WWI saw the introduction of metal helmets (see chapter 5), and in later conflicts fighters wore flak jackets.

Hence, the number of soldiers killed in battle dramatically understates the greater levels of violence in modern wars. For centuries, the ratio of wounded to killed hovered around three, and for the US military today that number is now closer to ten.[23]

Time, Distance, and Decision-making

Ancient and medieval armies plodded across continents for months, even years, waging prolonged sieges and sustained offensives. Today, the leaders of the United States, Russia, and China have to respond in minutes to an attack that could lead to a global war that is over in hours.

Geographical distance no longer protects civilian populations against the ravages of war. During our six million years as hunter-gatherers, most of us lived far apart, spread out across the globe. Even the fiercest band of hunter-gatherers could kill only a tiny fraction of the total human population. During more recent history, mountains, deserts, and oceans provided barriers to the armies of the world, limiting the damage militaries could inflict upon civilians. In the past, the practical logistics of feeding an army and the constraint of a reliable supply line bounded the projection of power. But today distance is no longer a defense. Projectiles can rocket through the skies at supersonic speeds to deliver atomic bombs.

For the first several million years of our existence, we lived in hunter-gatherer communities, in which the ability to exert control over others was limited. For the most part, no individual had weapons others did not have, and members of the tribe had similar roles and responsibilities. Hunter-gatherers lived in violent but egalitarian communities, without political hierarchies, in which the decision to go to war was made by those who fought. Decisions about war were made by the adults of the clan sitting around a fire.

During ancient times and the Middle Ages, a small fraction of the population determined the wars to be waged. Rulers fought to acquire other lands, to gain subjects, and to enslave people. These wars offered no benefit to soldiers, other than a return to subsistence farming after the fighting was done (if they survived). Slavery, indentured servitude, and similar institutions, enforced by the threat of violence, generated conscripts. Nevertheless, if large portions of an army refused to fight, the war was over. Until recently, political leaders were constrained by the willingness of at least a portion of the army to march into battle.

But modern technology has consolidated the power to wage war into the hands of just a few without the need to draw upon large armies. The president of the United States and the leader of Russia can single-handedly destroy most of the world in minutes, without the consent of the billions who will suffer the consequences. The ability to destroy entire nations on the other side of the planet in minutes is for the first time in the hands of one individual. Nevertheless, some take comfort in the fact that nuclear weapons have been with us for three-quarters of a century and yet we have somehow managed to avoid a global nuclear catastrophe. But this is another example of survivor bias distorting our reasoning.

At least twice we have come close to ending life on earth as we know it.

The 1983 Serpukhov-15 Incident: "It's a False Alarm"

At seven p.m. on the night of September 26, 1983, Lieutenant Colonel Stanislav Petrov began his twelve-hour shift in Serpukhov-15, an underground bunker outside Moscow. Serpukhov-15 contained an early warning system, known as the Eye, assembled by the Soviet Union to detect US nuclear missile launches. The Eye relied upon satellites specially designed to peer down on the United States in order to pick up heat trails from rocket launches and relay that information back to the bunker at Serpukhov-15. The commander of Serpukhov-15 would then notify the Kremlin leadership and launch a counterstrike. Given the curvature of the earth, a missile launch from the US could be detected by ground radar no more than fifteen minutes before reaching the Soviet Union. The satellites that fed the Eye system allowed the Soviets to double this notification period to about thirty minutes, allowing more time for Soviet missiles to escape their silos ahead of an incoming American attack.

Petrov was a software engineer by training and part of the team that had programmed the system. He normally worked in a building nearby. However, occasionally he was assigned a shift as commander of the bunker to check that the Eye's computers were functioning properly. Unlike others in the bunker, Petrov considered himself more a civilian engineer than a military officer. He was forty-four years old and had risen to deputy chief of combat algorithms for the Soviet Air Force.

Petrov sat in his chair above the main floor of the station looking down through a glass window to banks of telephones and electronic monitors staffed by Soviet Air Force officers. A large map covered the wall at the front of the room displaying the North Pole at the center with the United States and Soviet Union displayed on each side. The map was oriented to follow the projected path of US ballistic missiles north up over the Pole before turning south to strike at Moscow and other Soviet targets.

At twelve fifteen a.m., a siren wailed, shaking the glass in front of Petrov. He rose from his chair. A screen above the map at the front of the room flashed in bright red letters LAUNCH. This had never happened before. A small bulb lit up on the huge map on the wall at the front of the room and identified a US ICBM launch from Malmstrom Air Force Base in Montana. The military officers in the room below jumped out of their seats and began running around, grabbing phones, and shouting orders.

Petrov commanded them to sit down and begin a systems check. He wanted to be sure this was not some glitch in the program or a problem with the data from one of the satellites. Petrov also thought that launching a single missile was an odd way to start a war. Then, a siren wailed again and the screen above the map flashed a second time, indicating a second missile launch. Then a third, a fourth, and a fifth. The red letters on the panel were now flashing MISSILE ATTACK, the signal that the United States had launched a first strike.

On that September evening, tensions between the United States and the Soviet Union were running high. Earlier in the year, on March 8, 1983, President Ronald Reagan gave a speech to the National Association of Evangelicals in Florida in which he labeled the Soviet Union an "evil empire" and asked his audience to "pray for the salvation of all of those that live in that totalitarian darkness." He quoted with approval the words of a young father he had met: "I would rather see my little girls die now, still believing in God, than have them grow up under communism and one day die no longer believing in God." He called for those in the crowd to join "the struggle between right and wrong and good and evil."[24]

Equally troubling to the Soviets, Reagan was funding research for a defensive ballistic missile system known as Star Wars. If successful, the system would enable the United States to shoot down missiles launched by other countries. To the Soviets, this program suggested the United States believed it was possible to win a nuclear war by destroying missiles launched as part of a Soviet second strike. The Soviets warned the Americans that Star Wars would open a "new phase of the arms race" and that the nations were crossing a "dangerous red line" that risked global nuclear war.[25]

On September 1, just three weeks before alarms sounded in Serpukhov-15, Korean Air Lines Flight 007 from NYC to Seoul had been shot down by a Soviet fighter. A navigational error had led the plane into Soviet airspace, and Soviet radar misidentified the aircraft as US military. All 269 passengers and crew perished, including US Congressman Larry McDonald, a representative from Georgia. To make matters worse, the Soviets initially denied all knowledge of the incident, but then reversed position and admitted to the attack. The Soviets alleged the United States arranged for the plane to veer off course in order to probe Soviet air defenses.

Given the events of the previous months, Petrov knew that relations between the Soviet Union and the United States were deteriorating rapidly. A first strike by the United States was feared more than ever by the Soviet leadership.

Petrov picked up the handset cradled on the red phone next to his chair. The phone had only one purpose—to alert the Soviet leadership that the US had launched a first strike and wait for the orders to respond with a Soviet second strike. He spoke into the mouthpiece, "It is a false alarm."

And then he waited.

Those in the room below continued to frantically test and retest software programs and recalibrate satellite uplinks. There was no indication of a malfunction. Finally, after thirty minutes, despite the continued warnings and flashing lights, there were no explosions from above the bunker. It was clear there was no incoming US missile strike. Petrov relaxed back in his chair.

Petrov afterward stated that he believed if one of the regular military officers had been in command that September night, an incoming American missile strike would have been reported to Moscow and a counterstrike immediately ordered. Petrov typically served as commander of the bunker two nights a month.

It was later determined that sunlight bouncing off high-altitude clouds had created the false alarm. Petrov was initially told his actions were correct but was given no promotion or bonus and then later was reprimanded for failing to provide adequate documentation of his decisions. He was pushed out of the Soviet Air Force soon thereafter. After leaving the military, he worked at a research institute and lived in a small apartment in Moscow. In 2017, he died of pneumonia at age seventy-seven.

The Soviet and Russian governments never formally recognized Petrov's actions on that day, but he was given a World Citizen Award by the United Nations in 2006 and the Dresden Peace Prize in 2013. A documentary titled *The Man Who Saved the World* was released in 2014. In that film, when asked about the events on the night of September 26, 1983, Petrov looks into the camera and quietly says, "I was simply doing my job. I was the right person at the right place, that's all . . . I am not a hero."

Able Archer: Another Close Call

The events of that September night in 1983 are not the only time a false alarm has brought the world to the brink of global nuclear war.

In November 1983, just months after the crisis at Serpukhov-15, the Soviet Union prepared for a first strike by NATO forces.[26] Operation Able Archer, planned to commence on November 7, was a NATO exercise to practice responses to a nuclear war. But Russian intelligence believed it might be a cover by NATO for military preparations to destroy the Soviet Union.

In fact, Able Archer did simulate many of the actions that NATO would take before a first strike. NATO forces were moved from DEFCON 5 to DEFCON 1, the highest possible level. More than nineteen thousand soldiers were airlifted from the US to Europe. Command of the armed forces was shifted to the NATO Permanent War Headquarters. Two other factors compounded Soviet fears: evidence that the leaders of a number of NATO countries were directly participating in the exercise and intercepted encrypted communications that suggested a nuclear strike was imminent.

In response, the Soviet military placed military aircraft in Poland and East Germany on "strip alert," ready to take off immediately, armed with nuclear weapons. Soviet ICBMs were prepared for launch. The Soviet Army was put on a force-wide alert—for the first time since WWII.

Normally, such Soviet activity would have prompted NATO to respond in kind. But General Leonard Perroots, the head of US Air Force Intelligence, argued successfully to NATO commanders to de-escalate the crisis by not responding to heightened Soviet activity. Responding would have confirmed Soviet suspicions and likely would have led to further escalations. After Operation Able Archer was completed on November 11, Soviet forces were returned to regular status.

It is still unclear how close the world came to a nuclear war that November. Soviet and Russian officials have refused to discuss it. Many US government documents related to Able Archer have not been released for reasons of national security. Just before he retired in 1989, General Perroots wrote a letter to the White House about the events surrounding Operation Able Archer. Thirty-two years later that letter is still classified. Perroots passed away in 2017 at the age of eighty-three.

Not a Rare Occurrence

There have been other false alarms and near misses.[27]

- October 5, 1960. NORAD went on high alert after radar indicated a massive Soviet missile launch. The Soviet Premier Nikita Khrushchev was in New York at the time attending a UN meeting, and so the report was discounted. It later turned out that radar had mistaken a moonrise for missiles.

- May 23, 1967. The Air Force believed its Ballistic Missile Early Warning System, used to detect missile launches from the Soviet Union, was being jammed, which is considered an act of war. The US Air Force began preparing to launch a fleet of bombers. But soon, the weather forecasters at NORAD, who were providing information to the Air Force, realized the jamming was due to sun flares.

- November 5, 1979. National Security Advisor Zbigniew Brzezinski was awoken in the middle of the night to be told the Soviet Union had launched 250 missiles at the United States and President Carter needed to decide whether to retaliate in the next seven minutes. Before that time elapsed, he received another call that it was a false alarm. A training tape on how to respond to a nuclear attack had been inadvertently loaded into a US Air Force computer.

- January 25, 1995. Russian President Boris Yeltsin activated the Russia nuclear football after radar detected a missile that appeared to be from a US nuclear submarine stationed off the coast of Norway. But Russian satellites could not confirm the launch, so Yeltsin stood down. It turned out to be a Norwegian research rocket about which the Russians had been notified, but the information had not been passed up the chain of command.

Between 1962 and 2002, there were at least thirteen near-use nuclear US military incidents.[28] Marshall Shulman, a senior US State Department official, reviewed a number of these false alarms and in a top-secret brief, part of which has now been declassified, concluded that "false alerts of this kind are not a rare occurrence. There is complacency about handling them that disturbs me."[29] But even if political leaders act rationally and military officers are not misled by false alarms, nations can still blunder into battle, carried along by events beyond their control, stumbling forward in the fog of war.

In 1963, that is just about the way it happened.

The Cuban Missile Crisis

At seven p.m. EST, on Monday, October 22, 1962, President Kennedy went on national television to inform the nation that the Soviets were installing ballistic missiles in Cuba. The president announced a blockade of the island to begin on Wednesday morning and threatened that the launch of even a single missile from Cuba "against any nation in the Western Hemisphere" would result in "a full retaliatory response upon the Soviet Union."[30]

Kennedy was under tremendous political pressure to stand up to the Soviets. Midterm elections were only three weeks away and would be a public referendum on Kennedy's first two years as president. Just six days earlier, the one-page headline of *The New York Times* was "Eisenhower Calls President Weak on Foreign Policy." Republicans in Congress had spent the previous year calling the president feckless after the Bay of Pigs fiasco in April and the public relations disaster of the Khrushchev summit in Vienna in June.[31] After the summit, Vice President Lyndon Johnson told his friends, "Khrushchev scared the poor little fellow dead."[32]

From the start of the crisis, Kennedy was also under pressure from his own advisors and leaders in Congress to take immediate aggressive action against the Soviets. General Curtis LeMay told the president in the Oval Office that failing to attack the Soviets now was "as bad as the appeasement at Munich."[33] This was a particularly stunning remark, as Kennedy's father had been the ambassador to England before WWII and had been accused of harboring sympathies for the Nazis.[34] As commander of the US Air Force, LeMay's recommendation was "now that we have the Russian bear in a trap let's take his leg off right up to the testicles. On second thought, let's take off his testicles too."[35] LeMay was opposed to the blockade and instead proposed a nuclear strike and invasion. The chair of the Senate Armed Services Committee, Democratic Senator Richard Russell, told Kennedy that war with the Soviets was inevitable and should be fought now while the United States had more nuclear weapons than the Soviets.[36] Democratic Senator William Fulbright counseled an all-out invasion of Cuba "as quickly as possible."[37]

On the Soviet side, Khrushchev was gambling that Kennedy would not risk nuclear war over Soviet missiles stationed in Cuba. However, in the event of an invasion of the island, he was prepared to use nuclear weapons against American troops.[38] Khrushchev was sixty-eight years old, the son of a peasant farmer, and had fought his way to the top of the Communist Party. Kennedy was forty-five years old, the son of one of the richest men in America, and Kennedy's starter jobs had been serving as a US congressman and then US senator. Khrushchev thought he could take advantage of the young president he judged to be weak and inexperienced.

On Wednesday, October 24, the blockade of Cuba commenced with the US Navy surrounding the island. Over the next three days, Kennedy and Khrushchev stared each other down from thousands of miles away, waiting for the other guy to blink. Proposals and counterproposals were offered and rejected, threats were made and then withdrawn, and the world waited to find out if today was to be the last day of our species on earth.

On Saturday, October 27, the Soviet submarine B-59 suffered a breakdown in its ventilation system and submerged just outside the blockade.[39] Soon temperatures inside the sub reached as high as 140 degrees Fahrenheit, and carbon dioxide levels spiked. Some men collapsed and lay unconscious from heat exhaustion and carbon dioxide poisoning. Due to equipment failures, communications with the surface were lost. Around six thirty p.m., an American cruiser above the sub began dropping what the Soviet submarine commander thought were depth charges designed to sink the sub and kill all onboard. (They were in fact mock depth charges signaling the sub to surface.)

Exhausted and desperate, the sub commander, Valentin Savitsky, summoned the officer in charge of the nuclear torpedo to the engine room where it was housed. Arming the ten-kiloton nuclear weapon required action by both the officer in charge and the sub commander.

When the officer in charge arrived, Savitsky shouted: "We are going to blast them now! We will perish ourselves, but we will sink them all! We will not disgrace our Navy!"[40]

The officer in charge of the torpedo agreed but first wanted to consult with Vasily Arkhipov, the chief of staff of the submarine flotilla, who was of equal rank to Savitsky and happened to be onboard. Although not technically in the chain of command, Arkhipov argued against firing the weapon and managed to calm down Savitsky. At 9:52 p.m., the US Navy recorded that Soviet submarine B-59 surfaced, raised the Soviet flag, and steamed away. The sailors on the nearby American ships had no idea how close they had come to death.

If Arkhipov had happened to be onboard one of the other Soviet submarines, many believe B-59 would have launched its nuclear weapon.[41] Such an attack would have certainly destroyed the nearby aircraft carrier USS Randolph, its accompanying destroyers, and other US ships in the area, with a substantial loss of American lives.[42] United States intelligence at that time believed there were no nuclear weapons on Soviet subs and could have only concluded that the detonation was the result of a Soviet nuclear missile launched from Cuba on US forces.[43]

When B-59 returned to its home port in the Soviet Union, Arkhipov and Savitsky were criticized by Soviet officials for "violating the conditions of secrecy by surfacing" and were told they should have used the "special weapon."[44] Fortunately, history has treated Arkhipov more favorably than the reception he received in 1962 from Soviet authorities. In 2002, the US National Security Archive published an account of what took place on B-59 that day, crediting Arkhipov with "saving the world."[45] He died in Moscow in 1998, aged seventy-two.

Meanwhile, on that same Saturday, five thousand miles away over the skies of Russia, while the officers of Soviet sub B-59 were debating whether to launch a nuclear weapon, squadrons of Soviet and US fighters were racing toward each other, armed with nuclear missiles.

Earlier on that Saturday, Air Force Captain Charles Maultsby had been tasked to fly his U-2 reconnaissance plane over an area just south of the North Pole to monitor Soviet nuclear weapon testing.[46] Maultsby was an accomplished and experienced pilot and had done a stint flying acrobatics with the famous Air Force Thunderbirds. He was also a hardened military veteran who had been shot down during the Korean War and held as a prisoner of war for almost two years.

Unfortunately, Maultsby made several navigational errors and found himself three hundred miles off course over the Chukotka Peninsula of the Soviet Union in direct violation of Soviet airspace. Soviet MiG fighters were scrambled with orders to shoot down the intruder, which the Soviets believed could be the first of a fleet of nuclear bombers, a prelude to all-out war. Without communicating with Washington, the local American commander ordered US F-102 jets from Galena Air Station in Alaska to escort the U-2 back to base and defend the U-2. The F-102s were equipped with only air-to-air nuclear missiles.

Meanwhile, Maultsby had another problem besides the MiGs—he was running out of gas and was so far off course that he had no way to return to base in Alaska. Although he was still hundreds of miles inside the Soviet Union, he switched off his engines to save the last bit of fuel for landing and began slowly gliding away from Soviet airspace in the hopes of finding somewhere friendly to land.

At two p.m. EST, as Maultsby was silently slipping through the skies above Russia looking for a non-Russian airport, President Kennedy was in the Oval Office and informed of the situation. He was furious. Kennedy said, "There's always some sonofabitch that doesn't get the word."[47]

If the Soviet MiGs found the U-2, the US F-102s had orders to defend the spy plane with their nuclear missiles. But the Soviet MiGs had been chasing the U-2 for hours and, low on fuel, were forced to return to base. The US F-102s were able to guide Maultsby to a small, primitive ice landing strip just above the Arctic Circle. Later that day Khrushchev sent Kennedy a sternly worded message that the U-2 flight could have triggered a nuclear war.

The next day, Sunday, October 28, an agreement was reached between Kennedy and Khrushchev: The Soviets would return the missiles in Cuba to Russia in exchange for a US declaration not to invade the island. In addition, the two made a secret agreement that the US would dismantle its missiles in Turkey.

Looking back on the Cuban Missile Crisis, many believe the US and the Soviet Union were fortunate to have avoided an all-out nuclear war on October 27, the date that was described later by those in the White House as "Black Saturday." Dean Acheson, the former secretary of state, said years later that the peaceful outcome to the Cuban Missile Crisis was "plain dumb luck."[48] The secretary of defense during the Cuban Missile Crisis, Robert McNamara, concluded that "luck . . . played a significant role in the avoidance of nuclear war by a hair's breadth."[49] Future Secretary of Defense William Perry, who was an analyst at the time for the Department of Defense, said in 2015 that "the world avoided a nuclear holocaust as much by good luck as by good management." President Kennedy himself at the time stated the odds of global nuclear war during the crisis were between "one in three and even."[50]

With the help of some dumb luck, the United States and Russia have been able to maintain an uneasy nuclear peace. However, I believe the odds of a nuclear war will increase in the decades to come because there are likely to be more nuclear weapons in the hands of more nations. This is despite the laudable efforts of many to eliminate nuclear weapons from our planet.

Global Zero

Launched in 2008, Global Zero is an initiative to rid the earth of nuclear weapons. In 2009, US President Obama added his voice

Fred Schwed (1902–1966), author of *Where Are the Customer Yachts? or A Good Hard Look at Wall Street*, a wry and insightful commentary on those who work in financial services. *https://www.insider.co.uk/news/business-books-customers-yachts-fred-10135015*

Napoleon Hill (1883–1970), author of the best-selling book *Think and Grow Rich* which promised financial independence to those that followed his six-step method. *Archive PL. Alamy Stock Photo.*

Joseph Banks Rhine (1985–1980), author of numerous best-selling books on foretelling the future and promoter of branded playing cards to test for ESP. *Mary Evans Picture Library Ltd. Agefoto America.*

Robert Atkins (1930–2003), a medical doctor who sold over 45 million books on how to lose weight through "metabolic advantage." *Contributor, The LIFE Picture Collection. Getty Images.*

Theodor Sterling (1923–2005), a professor of computer science who was a paid researcher and spokesperson for the tobacco industry from the 1950s to the 1980s. *Simon Fraser University represented by the SFU Archives.*

A Frank Statement

to Cigarette Smokers

RECENT REPORTS on experiments with mice have given wide publicity to a theory that cigarette smoking is in some way linked with lung cancer in human beings.

Although conducted by doctors of professional standing, these experiments are not regarded as conclusive in the field of cancer research. However, we do not believe that any serious medical research, even though its results are inconclusive should be disregarded or lightly dismissed.

At the same time, we feel it is in the public interest to call attention to the fact that eminent doctors and research scientists have publicly questioned the claimed significance of these experiments.

Distinguished authorities point out:

1. That medical research of recent years indicates many possible causes of lung cancer.

2. That there is no agreement among the authorities regarding what the cause is.

3. That there is no proof that cigarette smoking is one of the causes.

4. That statistics purporting to link cigarette smoking with the disease could apply with equal force to any one of many other aspects of modern life. Indeed the validity of the statistics themselves is questioned by numerous scientists.

We accept an interest in people's health as a basic responsibility, paramount to every other consideration in our business.

We believe the products we make are not injurious to health.

We always have and always will cooperate closely with those whose task it is to safeguard the public health.

For more than 300 years tobacco has given solace, relaxation, and enjoyment to mankind. At one time or another during those years critics have held it responsible for practically every disease of the human body. One by one these charges have been abandoned for lack of evidence.

Regardless of the record of the past, the fact that cigarette smoking today should even be suspected as a cause of a serious disease is a matter of deep concern to us.

Many people have asked us what we are doing to meet the public's concern aroused by the recent reports. Here is the answer:

1. We are pledging aid and assistance to the research effort into all phases of tobacco use and health. This joint financial aid will of course be in addition to what is already being contributed by individual companies.

2. For this purpose we are establishing a joint industry group consisting initially of the undersigned. This group will be known as TOBACCO INDUSTRY RESEARCH COMMITTEE.

3. In charge of the research activities of the Committee will be a scientist of unimpeachable integrity and national repute. In addition there will be an Advisory Board of scientists disinterested in the cigarette industry. A group of distinguished men from medicine, science, and education will be invited to serve on this Board. These scientists will advise the Committee on its research activities.

This statement is being issued because we believe the people are entitled to know where we stand on this matter and what we intend to do about it.

TOBACCO INDUSTRY RESEARCH COMMITTEE

5400 EMPIRE STATE BUILDING, NEW YORK 1, N. Y.

SPONSORS:

THE AMERICAN TOBACCO COMPANY, INC.
Paul M. Hahn, President

BENSON & HEDGES
Joseph F. Cullman, Jr., President

BRIGHT BELT WAREHOUSE ASSOCIATION
F. S. Royster, President

BROWN & WILLIAMSON TOBACCO CORPORATION
Timothy V. Hartnett, President

BURLEY AUCTION WAREHOUSE ASSOCIATION
Albert Clay, President

BURLEY TOBACCO GROWERS COOPERATIVE ASSOCIATION
John W. Jones, President

LARUS & BROTHER COMPANY, INC.
W. T. Reed, Jr., President

P. LORILLARD COMPANY
Herbert A. Kent, Chairman

MARYLAND TOBACCO GROWERS ASSOCIATION
Samuel C. Linton, General Manager

PHILIP MORRIS & CO., LTD., INC.
O. Parker McComas, President

R. J. REYNOLDS TOBACCO COMPANY
E. A. Darr, President

STEPHANO BROTHERS, INC.
C. S. Stephano, D'Sc., Director of Research

TOBACCO ASSOCIATES, INC.
(An organization of flue-cured tobacco growers)
J. B. Hutson, President

UNITED STATES TOBACCO COMPANY
J. W. Peterson, President

On January 4, 1954, this full-page ad ran in over four hundred newspapers across the United States. The ad was paid for by the Tobacco Industry Research Committee, the industry group that funneled millions of dollars to Theodor Sperling and whose product was "doubt."

Abraham Wald (1902–1950), a math professor at Columbia University who as part of the Statistical Research Group identified survivor bias in research conducted by the US military during WWII. *University Archives, Rare Book & Manuscript Library, Columbia University Libraries.*

STATISTICAL RESEARCH GROUP, COLUMBIA UNIVERSITY

SECRET

2-28-5G

9.

Table 1

Losses and Distribution of Hits on Survivors
Inflicted by Enemy Fighters,
Eighth Fighter Command
24 August 1943 to 31 May 1944*

(1) Number of hits on survivors i	P-47			P-51		
	(2) Number of survivors with i hits R_i	(3) $(1) \times (2)$ iR_i	(4) $(1) \times (3)$ i^2R_i	(5) Number of survivors with i hits R_i	(6) $(1) \times (5)$ iR_i	(7) $(1) \times (6)$ i^2R_i
1	91	91	91	54	54	54
2	33	66	132	15	30	60
3	26	78	234	5	15	45
4	6	24	96	1	4	16
5	5	25	125			
6	4	24	144			
7	4	28	196			
8	1	8	64			
9	1	9	81			
10	2	20	200	1	10	100
11	1	11	121			
18	1	18	324			
30				1	30	900
Total	R = 175	T = 402	S = 1806	R = 77	T = 143	S = 1175
Losses	104			92		

* Data from "Battle Damage and Loss, VIII Fighter Command, 24 August 1943 - 31 May 1944," Operations Analysis Report, Memo M-41, VIII Fighter Command, 12 July 1944 (Secret). Thirty-seven aircraft returning with undetermined numbers of hits are distributed among the various numbers of hits in proportion to reported numbers.

SECRET

A now declassified secret report from the Statistical Research Group, a secret agency set-up during WWII within the Office of Scientific Research and Development, which reported to the President. This page is excerpted from Report 2-28-5G, written in 1945 by Abraham Wald summarizing his research on the vulnerability of Allied bombers to enemy fire. *National Archives.*

Shirō Ishii (1892–1959), creator and leader of Unit 731, the secretive group with the Japanese Army responsible for biological weapons programs from 1932 to 1945. *CPA Media Pte Ltd, Alamy Stock Photo.*

Remains of the main Unit 731 laboratory at Ping Fan, near Harbin, China, for experimentation on the effects of biological weapons on Chinese civilians and Allied prisoners of war. The Japanese military called this building the Epidemic Prevention and Water Purification Facility. *Xinhua News Agency, Getty Images.*

SUBJECT: Brief Summary of New Information About Japanese B.W. Activities

g. It was found that an organization completely separate from Pingfan had carried on a considerable amount of research in the veterinary B.W. field. At the present time 10 members of this group are engaged in preparing a report that will be available sometime in August.

h. General Ishii, the dominant figure in the B.W. program, is writing a treatise on the whole subject. This work will include his ideas about the strategical and tactical use of B.W. weapons, how these weapons should be used in various geographical areas, (particularly in cold climates), and a full description of his "ABEDO" theory about biological warfare. This treatise will represent a broad outline of General Ishii's 20-years' experience in the B.W. field and will be available about 15 July.

i. It was disclosed that there were available approximately 8,000 slides representing pathological sections derived from more than 200 human cases of disease caused by various B.W. agents. These had been concealed in temples and buried in the mountains of southern Japan. The pathologist who performed or directed all of this work is engaged at the present time in recovering this material, photomicrographing the slides, and preparing a complete report in English, with descriptions of the slides, laboratory protocols, and case histories. This report will be available about the end of August.

j. A collection of printed articles totaling about 600-pages covering the entire field of natural and artificial plague has been received; there is also on hand a printed bulletin of approximately 100 pages dealing with some phase of B.W. or C.W. warfare. These documents are both in Japanese and have not been translated.

3. The human subjects used at the laboratory and field experiments were said to be Manchurian coolies who had been condemned to death for various crimes. It was stated positively that no American or Russian prisoners of war had been used at any time (except that the blood of some American POW's had been checked for antibody content), and there is no evidence to indicate that this statement is untrue. The human subjects were used in exactly the same manner as other experimental animals, i.e., the minimum infectious and lethal dosage of various organisms was determined on them, they were immunized with various vaccines and then challenged with living organisms, and they were used as subjects during field trials of bacteria disseminated by bombs and sprays. These subjects also were used almost exclusively in the extensive work that was carried out with plague. The results obtained with human beings were somewhat

CONFIDENTIAL

Copy 1 of 4

A now declassified secret report prepared for the Chief of the Chemical Corps of the US Army, dated June 20, 1947, outlining the work Shirō Ishii and other former members of Unit 731 were conducting for the US government in exchange for immunity from prosecution.

THE WHITE HOUSE

WASHINGTON

August 11, 1945

My dear Mr. Cavert:

I appreciated very much your telegram of August ninth.

Nobody is more disturbed over the use of Atomic bombs than I am but I was greatly disturbed over the unwarranted attack by the Japanese on Pearl Harbor and their murder of our prisoners of war. The only language they seem to understand is the one we have been using to bombard them.

When you have to deal with a beast you have to treat him as a beast. It is most regrettable but nevertheless true.

Sincerely yours,

Harry Truman

Mr. Samuel McCrea Cavert
General Secretary
Federal Council of
    The Churches of Christ in America
New York City, New York

Letter from President Harry Truman to Samuel Cavert, head of the US Council of Churches. Cavert had sent a telegram to Truman after the bombing of Hiroshima asking Truman to refrain from the further use of atomic weapons. *Shapell Manuscript Foundation.*

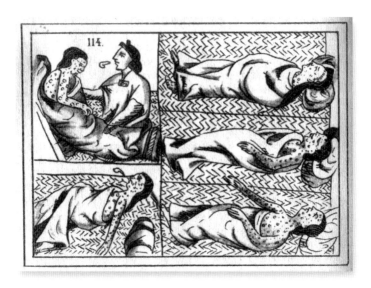

A 16th century Aztec illustration of smallpox victims.
The History Collection, Alamy Stock Photo.

William Whiston (1667–1752), one of the most well-known mathematicians and theologians of his day who was an early proponent of the theories of Isaac Newton as well as prophesies about the end of the world. *The Picture Art Collection, Alamy Stock Photo.*

A TRUE AND FAITHFUL

NARRATIVE

OF

WHAT PASSED IN LONDON,

DURING THE

GENERAL CONSTERNATION OF ALL RANKS AND
DEGREES OF MANKIND,

ON

TUESDAY, WEDNESDAY, THURSDAY, AND FRIDAY LAST.

On Tuesday the 13th of October, Mr. Whiston
held his lecture, near the Royal Exchange, to an
audience of fourteen worthy citizens, his subscribers
and constant hearers. Beside these, there were five
chance auditors for that night only, who had paid
their shillings apiece. I think myself obliged to be
very particular in this relation, lest my veracity should
be suspected; which makes me appeal to the men,
who were present; of which number, I myself was
one. Their names are,

> Henry Watson, haberdasher.
> George Hancock, druggist.
> John Lewis, drysalter.
> William Jones, cornchandler.

Henry

The front page of a satirical pamphlet written by Jonathan Swift, of *A Modest Proposal* and *Gulliver's Travels* fame, printed and distributed after Whiston's prediction of the end of the world on the evening of October 16, 1736, did not come to pass. *https://en.wikisource.org/wiki/Page:The_Works_of_the_Rev._Jonathan_Swift,_Volume_17.djvu/364*

Lieutenant Colonel Stanislav Petrov (1939–2017), the Soviet military officer who has been called the "Man Who Saved the World" on the night of September 26, 1983. *dpa Picture-Alliance. Agefoto America.*

A recent image of Serpukhov-15, where in 1983 Lieutenant Colonel Stanislav Petrov was stationed. Serpukhov is one of the early warning facilities that houses the "Eye" which watches for incoming ballistic missiles from the United States. The facility remains in operation today. *Imagery © 2021 CNES/Airbus, Maxar Technologies, Map data ©2021 Google.*

Vice Admiral Vasily Arkhipov (1926–1998), the Soviet submarine officer who convinced his comrades during the Cuban missile crisis to stop the launch of nuclear weapons to defend themselves against US military vessels dropping depth charges from the surface above. *https://commons.wikimedia.org/wiki/File:Vasili_Arkhipov_young.jpg*

Soviet submarine B-59 the day after "Black Saturday" with Vasily Arkhipov on board, steaming away from Cuba and back to the Soviet Union. A US Navy helicopter is hovering about tracking the sub. *U.S. National Archives, Still Pictures Branch, Record Group 428, Item 428-N-711200.*

Air Force Captain Charles Maultsby (1926–1998) whose navigational error took him over Soviet airspace during the Cuban Missile Crisis, nearly sparking a nuclear war between the Soviet Union and the United States. *US Air Force photo.* *https://nsarchive2.gwu.edu/nsa/cuba_mis_cri/dobbs/maultsby.htm*

Captain Maultsby's U-2 plane, serial number 56-715. This plane conducted the longest U-2 flight in history, eventually gliding to a landing on a secret US airstrip in the Arctic Circle. *US Air Force photo. https://nsarchive2.gwu.edu/nsa/cuba_mis_cri/dobbs/maultsby.htm*

John von Neumann (1903–1957), considered by many to be the "smartest man alive," who foresaw in the 1940s the twin dangers to the survival of the human species from atomic weapons and fossil fuels. *Everett Collection Historical, Alamy Stock Photo.*

Joseph Fourier (1768–1830), the French mathematician and friend of Napoleon who discovered what others would later call the "greenhouse effect." *https://commons.wikimedia.org/wiki/File:Fourier2.jpg*

The cartoon by Alan Dunn published in *The New Yorker* on May 20, 1950 portraying aliens stealing New York City trash cans. This drawing prompted Enrico Fermi to question the existence of intelligent life in the universe. *Alan Dunn, The New Yorker Collection/ The Cartoon Bank.*

Enrico Fermi (1901–1954), a Nobel Prize-winning physicist who asked the question concerning aliens, "Where are they?" or what has become known as Fermi's Paradox. *The Print Collector, Alamy Stock Photo.*

Erwin Schrodinger (1887–1961), a Nobel Prize-winning physicist who co-founded quantum mechanics and originated the paradox of Schrodinger's Cat from a thought experiment suggested by Albert Einstein. *Pictorial Press Ltd., Alamy Stock Photo.*

to those of other world leaders in a speech in Prague in support of Global Zero.

But even if the proponents of Global Zero could snap their fingers and rid the world of nuclear weapons tomorrow, they would have simply replaced the old nuclear arms race with a new race to build nuclear arms. All major nations would still have the ability to rapidly produce and deploy nuclear weapons. All countries would be subject to the risk that the first nation to rebuild its nuclear arsenal could threaten the others with annihilation. As a consequence, all countries would put in place the necessary staff and materials to rapidly rebuild a nuclear arsenal in case the other nations did the same. It would be impossible to verify and to police the potential capacity of other nations to build nuclear weapons. Such infrastructure could be easily hidden among other manufacturing or industrial complexes.

A race to build nuclear weapons could be more destabilizing than a nuclear arms race. In a world without nuclear weapons, any country that suddenly started to rebuild its nuclear capability would force all other countries to follow. During that period, there would be an incentive to use conventional weapons to strike at the country that started reconstructing atomic bombs. Furthermore, if a country were to gain nuclear capability before others, it might consider nuclear war winnable. A world with a single nuclear power, even for a short period of time, would be a frightening place, particularly if that country were a rogue or terrorist state. Imagine if a North Korea or a Syria were the only nation in the world with nuclear weapons. Unfortunately, bad actors have the most incentive to cheat in a Global Zero world. And Global Zero would likely not really be zero. As a practical matter, most nations would probably hide a few nuclear weapons, just in case others did the same.

No First Use

Besides reducing the number of nuclear weapons, another proposal to reduce the risk of nuclear war is for nations to adopt a no first use (NFU) policy. The logic of NFU is that if no nation launched a first strike, there could be no nuclear war. Another benefit of NFU is that the number of nuclear weapons deployed by a country could be reduced in number. Because a first strike capability requires sufficient warheads to take out all or most of the defending country's nuclear weapons, those countries fearing a first strike must maintain constant readiness. But if all nations committed to NFU, a mutual reduction in missiles could be achieved, as an equal number of missiles on all sides provides a sufficient deterrent.

Today, only China and India have publicly committed to NFU, although India has carved out exceptions for responding to chemical and biological attacks and for "punitive retaliation should deterrence fail." (In other words, they retain the option of using nuclear weapons against Pakistan.)[51] The United States has never committed to NFU. In fact, the opposite has been part of United States military strategy for the past seven decades.

After WWII, in its battle against global communism, the United States made a public commitment to defend the NATO countries, Japan, South Korea, and Taiwan. This pact is more formally known as the Positive Security Assurance (PSA).[52] Part of the reason the US put in place the PSA is that stationing conventional forces in those countries sufficient to repel an attack by the Soviets or China would be prohibitively expensive. Instead, the United States covers those nations with a nuclear umbrella that promises a nuclear response to a conventional attack. Because nuclear weapons are a lot less costly than conventional forces, particularly those stationed overseas, this policy has saved the American taxpayers hundreds of billions of dollars over the last seventy years.

This savings can be seen in defense expenditures. United States spending on the military was less than 10 percent of GDP between the Korean and Vietnam Wars, declined to 6 percent during the Reagan years, and is 3 percent today.[53] By contrast, the Soviet Union consistently spent 25 percent or more of GDP on defense, and this burden bankrupted the country.[54] More than two hundred Soviet Army divisions were in Europe during the peak of the Cold War, compared with fewer than twenty US Army divisions.[55]

For the United States, nuclear weapons have been a way to purchase deterrence on the cheap. And the US nuclear umbrella conserved more than just taxpayer money. Without nuclear weapons, the United States would have had to reinstitute the draft, taking men and women out of the workplace and home and placing them into military service overseas. Since the 1970s, no American has been forced to serve or leave their country, which has been a policy quite popular with sons, daughters, and parents alike.

Despite the coverage the nuclear umbrella offered to our Allies in the past, some question whether the PSA is still in force as a practical matter. Would a US president today risk Chicago for Seoul? Atlanta for Warsaw? Committing to NFU would mean publicly withdrawing assurances that the US would respond to an invasion of a US ally by Russian or Chinese conventional forces with a nuclear strike. Without the threat of a nuclear strike, the allies that rely upon the PSA are largely defenseless—unless the US were to deploy hundreds of thousands of American troops overseas.

If the United States were to publicly adopt an NFU policy, then many of our allies would have no choice but to develop nuclear weapons for self-defense. Neither Japan nor South Korea could afford a conventional force sufficient to repel the Chinese Navy, Army, and Air Force. Nuclear weapons would be the only alternative. The same calculation applies to NATO countries neighboring Russia. Without US guarantees, nuclear weapons could proliferate to other nations throughout the world, including longstanding enemies. Parts of the world that are nuclear-free today would bristle with missiles topped with atomic bombs in armed camps. If the United States adopted an NFU policy, the nuclear club could have a dozen new members.

In the years to come, I think American leaders will acknowledge what some believe today to be the United States' de facto nuclear policy: NFU except for attacks of any type on US soil. Even if that public admission never comes, many countries will increasingly question whether the United States will run to their defense at the risk of a global nuclear war. And that will lead them to develop nuclear weapons of their own.

Measures to Reduce the Risk of Nuclear War

For the reasons outlined previously, I believe the proliferation of nuclear weapons has just begun. There are currently nine members of the nuclear club: the United States, Russia, China, India, Pakistan, United Kingdom, France, Israel, and North Korea. More countries are likely to join, all unwelcome and some particularly unwelcome, such as those in the Middle East.

While I do not foresee a way to prevent this, I believe there are two specific steps that could be taken unilaterally by the United States to reduce the threat of global nuclear war.

Consensus of Five

The president of the United States, as commander in chief, has the unilateral right to exterminate billions in minutes. We should not place in the hands of one person the power to end life on earth as we know it. (It would appear that some other nuclear powers have also delegated the power to make this decision to one individual.) There is nothing in the US Constitution or federal laws that grants this authority solely to the president—it has just been the practice since Truman. It is a practice that could be easily changed.

Before pushing the nuclear button, we should require that a majority of the president, vice president, secretary of defense, speaker of the house, and the president pro tempore of the Senate have to agree to end life on earth as we know it.

No First Day

Stated US policy is that the president may "launch on warning" if there is a report of an incoming first strike. The rationale is that the US commander in chief has to respond in kind before the attacker destroys the US nuclear arsenal, rendering the nation defenseless. In other words, "use 'em or lose 'em." The time between receipt of the warning of a first strike and when the president has to make this decision is only minutes.

The US has approximately six thousand nuclear weapons topping missiles burrowed into hardened silos across the country, loaded on hundreds of military aircraft at the ready, and hidden in the holds of about two dozen submarines lurking beneath the surface of the oceans. Even if an attacker could wipe out 90 percent of these weapons, which is highly unlikely, the remaining six hundred atomic bombs would be more than sufficient to launch an annihilating counterstrike.

Hence, we should demand the public adoption of a No First Day (NFD) policy by the United States. Under this policy, the US would agree to wait at least twenty-four hours before responding to a first strike. The US should then encourage other countries to adopt a similar policy.

The US nuclear arsenal is today on a hair-trigger. The stated policy of "use 'em or lose 'em" is highly destabilizing: National leaders are expected to launch on warning, which risks an all-out global exchange of nuclear weapons due to a false alarm, misinterpreted signal intercept, failure in the chain of command, etc. An NFD policy removes the pressure to decide the fate of humanity in minutes, a decision that is worth reflecting on overnight.

If the US adopted this policy, Russia, China, and other nuclear nations may follow—though probably only if the US took the lead. But once the US did so, there would be less incentive for others to be on such high alert. If the purpose of nuclear weapons is deterrence, then there is no reason for thousands of them to be locked and loaded, ready to be fired at a moment's notice.

I am not proposing the nations of the world surrender all or even some of their weapons—that is unrealistic. But I don't see why we couldn't all agree to put down our guns.

Even Ronald Reagan, an avowed Cold Warrior, wrote in his autobiography:

> *The decision to launch the (nuclear) weapons was mine alone to make . . . Six minutes to decide how to respond to a blip on a radar scope and decide whether to unleash Armageddon! How could anyone apply reason at a time like that . . . how do we go about trying to prevent it and pulling back from this hair-trigger existence?*[56]

Conclusions

As we have seen, seventy years of living with tens of thousands of nuclear weapons, poised to be launched at a moment's notice, has almost ended all life on earth as we know it more than once. If not for "dumb luck," then at least twice, on the night of September 26, 1983, and on Black Saturday during the Cuban Missile Crisis, the United States and Soviet Union would have fought a nuclear war. When contemplating the years to come, we should acknowledge that the fact we have survived the last seven decades without a nuclear exchange between nations was an unlikely outcome.

Furthermore, the odds of avoiding a global nuclear war in the future seem to be increasingly stacked against us. An announced or de facto withdrawal of the US and Soviet protective nuclear umbrellas would push some nations to develop nuclear weapons to defend themselves against larger conventional forces and other neighboring countries. Hence, the nuclear club will likely have some new members from some of the most unstable parts of world.

There are those who take comfort in the fact that only two atomic bombs have been exploded in anger in human history. So far, the world has found a way since WWII to maintain an uneasy nuclear peace. But as seen, we have been more lucky than good. Next time a Stanislav Petrov or a Vasily Arkhipov might not be there. Or maybe on a future Black Saturday the MiGs don't run out of gas.

The philosopher Bertrand Russell wrote about nuclear war: "You may reasonably expect a man to walk a tightrope safely for ten minutes; it would be unreasonable to do so without accident for two hundred years."[57]

Another threat to our future survival is global warming. Our reliance on fossil fuels risks turning large swaths of our lush, green planet into burning deserts. The mechanisms that are causing today's global warming were first identified by a Frenchman over two centuries ago. This Frenchman was committed to science. If he had had more time to devote to his chosen profession, who knows what other discoveries he could have made besides the greenhouse effect.

But Napoleon had other ideas.

CHAPTER 9

Climate Change: Hot and Parched

Joseph Fourier and Glass Boxes

Joseph Fourier (1768–1830) is best known for the Fourier series, a method to sum a series of waves into one equation.[1] The Fourier series, and what is called Fourier analysis, can be used to solve thorny problems in heat transfer, harmonics, and signal processing.

Fourier was born to a poor family in Burgundy, in what is now France. His mother died when he was nine and his father died a year later, so he was sent to the local Benedictine abbey to study for the priesthood. In his twenties, Fourier decided on a career in politics rather than the Church and joined a local Revolutionary Committee of the French Revolution. But his faction within the Terror soon fell out of power. Fourier was accused of treason, convicted, and sent to prison. Before facing the guillotine, his faction came back into favor and he was released.

After his near-death experience, Fourier decided that a career in politics might not be for him. Instead, he became a student of the famous French mathematician Joseph-Louis Lagrange. When Lagrange retired, Fourier succeeded him as chair of mathematics at École Polytechnique.

Fourier's work at the Polytechnique attracted the attention of Napoleon Bonaparte. The Little Corporal was so impressed with the young mathematician that in 1798 he informed Fourier that his new job was scientific advisor to the French Army. Napoleon then promptly set off to Egypt with Fourier in tow, much to the mathematician's dismay. Napoleon grew to like Fourier and subsequently appointed him governor of Lower Egypt. When Napoleon returned to Paris from Egypt, he assigned Fourier to be governor of northern Italy.

In 1814, after Napoleon's first defeat, Fourier pledged allegiance to the king in order to return to Paris and his academic life. This proved to be the wrong decision, as Napoleon returned to power in 1815 and promptly placed Fourier in prison for treason. Fearing another encounter with the guillotine, Fourier renounced his pledge to the king and reaffirmed his fealty to Napoleon.

Napoleon was so happy that his former friend had repented his Royalist ways that he released Fourier from prison and awarded him the Prefecture of Rhône, complete with a lifetime annual pension of 6,000 francs. The appointment was effective July 1, 1815. However, Napoleon abdicated on June 22 in favor of his son and fled France on June 29. The new French government immediately withdrew the appointment and voided Fourier's pension. But this allowed Fourier, after almost two decades of trying to get out of government, to return to the safety of academics. He joined the French Academy of Sciences, away from the vicissitudes of French politics, where he remained until his death.

The Greenhouse Effect:
Why the Earth Is Not a Snowball

Fourier's discovery of the greenhouse effect derived from his work on heat transfer. Fourier understood that when heat energy strikes an object, some of that energy is absorbed and some is reradiated. If the object struck is at a lower temperature than the source of the heat, the energy is reradiated at a lower frequency. Conversely, if the object struck is at a higher temperature than the source of the heat, the energy is reradiated at a higher frequency.

In the case of the earth and the sun, the earth is at a lower temperature. When the sun's rays, which are in the visible higher frequency section of the light spectrum, strike the earth, some of that energy is absorbed, warming the land and oceans, and some is reradiated, in the invisible lower frequency part of the light spectrum. That is why we don't see the sun's reradiated rays returning to the skies above.

Based on his theories of heat transfer, Fourier calculated that, given the relative rates of absorption and reradiation of sunlight, the earth should have a global average temperature significantly lower than it has, or about zero degrees Celsius. This finding led him to conclude that the earth's atmosphere must be warming the planet. The analogy he used referenced a well-known fact at the time: If a box with a glass cover is placed in the sun, the air inside will rise in temperature. (Later, others would incorrectly label this the greenhouse effect. Greenhouses and boxes with glass covers warm air because the temperature of the box itself rises, and the glass prevents the air warmed by the box from escaping. But the earth does not have a glass cover, and heat from the earth's surface rises into space.) In fact, the warming of the earth's atmosphere is due to the rate at which water molecules and greenhouse gases absorb energy of different frequencies.[2] The chemistry is that water molecules and greenhouse gases are better at absorbing lower frequency radiation. The atmosphere, which is filled with water molecules and greenhouse gases, absorbs more lower frequency reradiated heat from the cooler earth than higher frequency heat from the hotter sun. Thus, the more greenhouse gases and water molecules that are in the atmosphere, the warmer the planet. The water molecules and greenhouse gases in the atmosphere let the sun's rays check in, but not check out.

We need look no further than our neighboring planet of Venus to find an example of what happens if the greenhouse effect spirals out of control.

Venus is about the same size as Earth, just six hundred fewer kilometers in diameter, and 80 percent of the mass. Our neighboring planet at one time contained an atmosphere not that different from that of the early Earth and might have been habitable several billion years ago.[3] However, Venus is much closer to the sun than we are, so the amount of energy that falls on Venus is about twice that of Earth. The resulting higher surface temperature over time caused a higher rate of evaporation of the water into the atmosphere. This additional water in the atmosphere trapped the reradiated heat from sunlight and warmed the planet further, which in turn caused an even higher rate of evaporation and more heat. At some point, this cycle of ever greater evaporation and heat reached a tipping point, and the temperature on Venus spiraled upward. This is known as a runaway greenhouse effect. Eventually, the temperature on Venus stabilized at about 525 degrees Celsius.

If Venus were a steak, it would be well done.

Feedback Loops on Earth

The evaporation of water into the atmosphere on Venus is an example of a positive feedback loop. A positive feedback loop spirals: Increases in the first variable (e.g., water in the atmosphere) cause an increase in the second variable (e.g., the temperature of Venus) that loops back to increase the first variable, and so on. A positive feedback loop spirals until a new equilibrium is reached. On the other hand, a negative feedback loop stabilizes: Increases in the first variable result in a decrease in the second variable, and so the system returns to the equilibrium of the initial state. (This is confusing, as negative is actually positive, as we want more negative feedback loops in the Earth's atmosphere. Maybe we should come up with new names.)

In terms of feedback loops, Earth's climate is more complicated and unpredictable than Venus's climate. This is because Earth's atmosphere has multiple positive and negative feedback loops. Each of these feedback loops interacts not only with itself but also with others. Normally we could apply the scientific method—test, retest, and look at the results—to determine how this complicated system works. But there is no other Earth or comparable nearby planet on which to experiment. So, there is no practical way, other than risking a global disaster from a runaway greenhouse effect, to determine by experimentation how our climate will react to changes in the composition of the atmosphere. That said, climate scientists have identified the important positive and negative feedback loops curling throughout Earth's atmosphere.

The most important positive feedback loop on Earth has already been discussed: More greenhouse gases in the atmosphere cause higher temperatures, resulting in more water evaporation. To get a sense of the magnitude of this positive feedback mechanism, 18 billion tons of water evaporate from the ocean into the atmosphere every second.[4] Another important positive feedback loop is the amount of ice on land and sea. Lighter objects reflect more heat and thus absorb less energy. White ice reflects 85 percent of sunlight compared with just 5 percent for the darker-colored land and oceans.[5] (That is why we wear lighter-colored clothes in the summer.) As Earth warms, ice melts and gives way to darker-colored surfaces, which retain more heat. An additional important positive feedback mechanism is from methane, the second most prevalent greenhouse gas after CO_2. Methane is released by the melting of the permafrost in the high latitudes of the planet.[6] Another positive feedback loop is the ocean's ability to absorb CO_2. The ocean is one large "heat sink" that pulls CO_2 out of the air and thus the more saturated the ocean becomes with CO_2, the less CO_2 it can absorb.[7]

The main negative feedback loop on Earth is rain. Rainfall is sensitive to temperature—the hotter it gets the more it rains. Rain catches CO_2 in the air and drags it down to Earth's surface to be deposited in rocks. An additional but less important negative feedback loop is photosynthesis. Plants use the sun's energy to combine CO_2 with water to produce more plants. More CO_2 in the atmosphere spurs more plant growth, and more CO_2 is then taken out of the atmosphere.

Earth and the Runaway Greenhouse Effect

How these positive and negative feedback loops react and interreact is difficult to quantify. But we do know what the net effect to date has been: The amount of CO_2 in the atmosphere has gone from 280 parts per million in 1750 to over 400 parts per million today, and global temperatures have risen one degree Celsius during the same period.[8] Intentionally or not, we have run an experiment on the global climate over the past 271 years and the results are in: Adding more CO_2 to the atmosphere raises global temperatures.

For our planet to meet a fate similar to that of Venus, it would require that at least one of the positive feedback loops reaches a tipping point that overwhelms all the negative feedback loops, triggering a runaway greenhouse effect. What it would take for humans to put enough CO_2 into the atmosphere to reach that tipping point and when that would happen is a subject of debate within the scientific community.[9] However, given survivor bias, the fact that we have not crossed that tipping point should not give us a lot of comfort.

What is certain is that if we continue to increase the amount of greenhouse gases released into the atmosphere each year, global temperatures will rise. And every year in which we add more CO_2 to the atmosphere increases the odds that global warming could spiral out of control.

Stopping Global Warming: Two Fixes

If we want to continue to call Earth home, there are two ways to stop global warming: One is to curb greenhouse emissions, and the other is to geoengineer the planet.

The amount of greenhouse gases released into the atmosphere is basically a function of the number of *Homo sapiens* on Earth times the amount emitted per person. Both have been rising steadily since the start of the Industrial Revolution over two centuries ago.

The good news, at least in terms of global warming, is that the number of people on the planet is projected to top out at about nine billion by the end of this century as the world becomes older and richer and birth rates steadily decline.[10] That is fortunate since the rate of population growth during the past century is unsustainable, not just for greenhouse gas emissions but also for food supplies and natural resources. The global population rose from 1.6 billion in 1900 to 6 billion in 2000. At that rate, the number of people on planet Earth would reach 22.5 billion by 2100. And even at a much slower rate of growth, an increase from seven billion today to nine billion by 2100 represents a 29 percent increase in the world population. If we wish to halt the growth of greenhouse gases, that means a 29 percent decrease in the amount emitted per person during the balance of this century.

The bad news is that the amount emitted per person is almost certain to rise. Improvements in the standards of living in the developed world have increased the amount of energy consumed per person over the past two hundred years. During the nineteenth and twentieth centuries, the estimated amount of energy consumed per capita increased five times and sixteen times, respectively.[11] That statistic is not surprising, as the Industrial Revolution has increased global economic output by more than one hundred times.[12] This trend is not expected to reverse. In fact, at the start of the twenty-first century, energy consumption per capita increased in the developed world as a whole, despite recent declines in the United States, and continues to rise. But this is not the real problem.

Today, about one billion people, those living in the developed nations of the West, emit most of the world's greenhouse gases. If a sizeable portion of the remaining six billion people adopt a Western lifestyle, then the amount of greenhouse gases emitted per person will skyrocket. For example, greenhouse gas emissions per capita in the United States is currently 17.3 tons. The same number for India is 1.4.[13]

In 1928, Mahatma Gandhi prophetically said: "God forbid that India should ever take to industrialism after the manner of the West . . . if [our nation] took to similar economic exploitation, it would strip the world bare like locusts."[14]

India is not the only nation whose emissions per capita are expected to rise. In China, greenhouse gas emissions per capita are 5.4 tons and growing rapidly as the Chinese people aspire to a Western standard of living.[15] In Africa, greenhouse gas emissions per capita are a low 0.9 tons, and that continent's population is the fastest growing on earth.[16]

There have been efforts to curb greenhouse gas emissions. In 1997, world leaders convened in Kyoto and committed to reducing reliance on fossil fuels. At the time of the conference, fossil fuels supplied 80 percent of the world's energy demands.[17] More than twenty years later, that number is 81 percent.[18] In 2016, world leaders met again, this time in Paris, and agreed to limit global average temperature rises to two degrees Celsius above preindustrial levels. The largest emitter of greenhouse gas, China, agreed to "level off" its emissions by 2030, but did not agree to limit emissions before then. India, another large emitter, agreed to limit its emissions, but also not before 2030. The United States committed to cut its greenhouse emissions by a quarter below 2005 levels by 2025, and then less than a year later pulled out of the Paris Agreement, only to rejoin in 2021.

Despite lots of summits, speeches, and pledges by world leaders, non–fossil fuels have increased from 14 percent of global energy consumption in 1971 to just 18 percent in 2014.[19] Progress toward renewables in 2018 was 85 percent short of the Paris Agreement goals.[20] Not a single major country is currently on track to meet its pledges.[21]

This is not a book to review the economics of solar, wind, geothermal, biomass, nuclear, and fossil fuels. But there are such books, and they reach the same conclusion: Renewables are more expensive than fossil fuels in most places. Regardless of the relative environmental merits and economic costs, studies show that few countries are implementing renewable policies to meet the commitments mandated by the Paris Agreement.[22] Because fossil fuels are less expensive, they remain the dominant source of energy on our planet.

Besides economics, there is an equally difficult problem with renewables—the nature of international climate agreements. The Paris Agreement carries no penalties, and so compliance is purely voluntary. The cost of noncompliance is that a country in the future will be accused of breaking a promise made by a politician in the past. With no meaningful penalty, the easy path for politicians is to promise now and apologize later. In addition, there is an incentive to cheat on Paris Agreement commitments because greenhouse gas emissions are a classic "tragedy of the commons."

Climate Agreements and Overgrazing the Atmosphere

The term "tragedy of the commons" arose in nineteenth-century England to describe farmers' behavior on land that was collectively owned. Farmers sent all their cows to graze on the commons, overgrazing killed all the grass, and the lack of grass killed all the cows. At the end, everyone was worse off, particularly the cows, but each individual farmer had acted rationally, given that all the other farmers were also sending their cows to graze on the same land. In the analogy of the tragedy of the commons, carbon emissions are the grass, the atmosphere is the commons, and we are the cows.

The long lead time to make the switch from fossil fuels to renewables also provides an incentive to "overgraze." Energy infrastructure demands decades of sustained investment. Building a nuclear power plant takes ten years or longer. Wind and solar farms are multidecade commitments. If a country makes such an investment, but then other countries do not make similar investments, those countries that did not invest in renewables gain the advantages of a reduction in greenhouse gases without incurring the costs of more expensive renewable energy.

Domestic Politics and Climate Agreements

The Paris Agreement stipulated that countries invest in renewables now to cut greenhouse gas emissions in the future. But within a year of signing the papers, the United States withdrew from the agreement. It is politically difficult for other countries to put aside money for initiatives that depend for success on multidecade commitments by all countries when one of the largest emitters on the planet pulls out soon after signing. While America did rejoin in 2021, other countries must wonder whether we will see another withdrawal after a subsequent US presidential election. Climate agreements don't work if there are opt-out clauses every four years.

Furthermore, domestic politics in the developing countries do not favor joining global climate agreements. The Western developed nations have enjoyed substantially higher standards of living due to industrialization for over two centuries. Seven percent of the global population has been responsible for more than half of all greenhouse gas emissions to date.[23] The cumulative emissions of greenhouse gases from the United States are more than three times that of China.[24] The developing world, particularly China and India, have only recently transitioned from agricultural to industrial economies. The problem of greenhouse gases in the atmosphere is largely due to the relatively luxurious lifestyle of those in the Western developed nations.

It is true that China emits more greenhouse gases today than the United States. But that's because China has more people. Greenhouse gas emissions per capita are 17.3 and 5.4 tons for the United States and China, respectively.[25] Based on these numbers, many in China question whether they should cut back on greenhouse emissions at all. They argue that the West is largely responsible for this mess, due to reckless and excessive energy consumption in years past that continues to this day. In other words, the West broke it, so let them fix it.

Some in China have proposed a cap on per capita greenhouse emissions for all countries. The argument is that all nations should be held to the same standard: Human life is of equal value regardless of where an individual was born. Under this proposal, that number would be set at the current global average—4.5 tons per capita—a logical place to start if the objective is to stop further increases in CO_2 emissions. Needless to say, proposals for the citizens of the developed West to slash their greenhouse gas emissions dramatically will not go far. A cap of 4.5 tons per capita would represent a 70 percent cut in emissions for the average citizen in the United States.

Intergenerational Issues: Great-Great-Great-Grandchildren

Some argue the costs of reducing greenhouse gas emissions should be paid for by future generations.

In most countries in the world, the government redistributes wealth from the rich to the poor. This is considered to be just and fair. To be consistent, this same standard should be applied between generations—a human life in the future should be no less valued than a human life today. Given expected advances in technology, due to the hard work and investments by the current generation, future generations will enjoy a much higher standard of living. Future generations will be richer, by any measure, than we are. And not just economically. It is probable that many diseases, such as cancer, diabetes, and strokes, will eventually be cured. Due to the efforts of those alive today, future generations will live longer and be healthier. Therefore, it is argued that the costs of fixing global warming should be borne by the relatively rich generations of the future—not the relatively poor generation of the present. (This same argument could apply to government deficits: The trillions in debt now being racked up each year are simply an intergenerational welfare program paid for by our wealthy, generous descendants.)

The Paris Agreement relies on a significant amount of intergenerational altruism. This has yet to be evidenced by any large nation. The tug on the heartstrings of most parents to make a better life for their children is strong. That same pull for the great-great-great-grandchildren is less forceful. There is also a more practical political reality: Intergenerational transfers of wealth from future to present generations are quite popular with elected officials. Individuals from whom the monies are taken, those not yet born, are a constituency that offers neither money nor votes in the next election.

Given the preceding economic, political, and intergenerational dynamics, I believe that the most likely case is that greenhouse gas emissions will continue to rise. Without immediate significant reductions in emissions, by 2100 the global temperatures could rise above pre–Industrial Revolution levels by as much as seven degrees.[26] There is debate within the scientific community over this number. But there is widespread agreement that we risk a significant rise in global temperatures if we continue on our current course and speed.[27]

Tragically, nothing is likely to happen, in my view, until there is a global crisis. At that point, our only option will be geoengineering.

Geoengineering: Barking Mad

There are two basic types of geoengineering: carbon dioxide removal (CDR) and solar radiation management (SRM). The first removes greenhouse gas from the atmosphere, and the second deflects sunlight away from the earth.

CDR, sometimes referred to as carbon sequestration, involves measures that suck CO_2 or methane out of the atmosphere, such as direct carbon capture, biochar, or ocean fertilization. The most promising approach is direct carbon capture, which removes CO_2 by passing air through a filter and then compressing the separated CO_2 gas into pellets to be stored underground. A leading company in this area, Carbon Engineering of Canada, promises that over the next ten years it will develop technology that cuts the cost of removing carbon from the air to potentially as low as $100 per ton, down from the current $600 per ton.[28] Global emissions of

CO_2 are about 35 billion tons per year.[29] If $100 per ton could be achieved, then to offset a 10 percent increase in CO_2 emissions would cost $350 billion dollars annually. At current prices of $600 per ton, the cost would be $2.1 trillion per year. Furthermore, the captured carbon has to be somehow stored safely underground, either injected into deep wells or buried in solid form. But leaks could contaminate water supplies, and any contact with high concentrations of CO_2 is toxic to humans. In 2010, plans to build seventy-seven CDR plants were announced. In 2019, that number was forty-five. The forty-five plants represent a capacity of 80 million tons per year, or about 0.2 percent of global CO_2 emissions.[30] As experts have recently concluded, "The forecast is still not bright for carbon capture and storage."[31]

Given CDR's costs and other complexities, SRM is most commonly talked about as a practical solution. SRM is cost effective and could reverse global temperatures in short order.

The basic idea of SRM is to reflect a portion of the sun's rays back into space before they can enter the earth's lower atmosphere. This already happens to some extent: 30 percent of sunlight is reflected away from the planet by clouds and other particles in the atmosphere.[32] SRM works by injecting sulfate particles, basically the equivalent of dust, into the upper atmosphere. Rain does wash these particles out of the sky, so continuous spraying from a fleet of specially equipped planes would be necessary. A study estimated that constructing a fleet of specially designed airplanes for spraying would cost $7 billion up front and $2 billion per year to operate.[33] Continuous spraying of eighteen thousand gallons per hour into the atmosphere would be sufficient for the entire planet.[34] These amounts of sulfate are relatively small in comparison to a big sky because we would only have to scatter about 2 percent of the sun's rays to reverse the effect of human-related greenhouse emissions to date.[35] Because SRM is so effective, it could rapidly cool the planet back to preindustrial levels and keep it there.

And herein lies the problem. A relatively small amount of sulfate makes a big difference in global temperatures. We could miscalculate the quantity of particles required. And we don't know exactly how changing the amount of sunlight entering the atmosphere would interact with the earth's multiple positive and negative feedback loops. To calculate how much sulfate to spray, researchers have developed complex models of the earth's climate with thousands of variables populating formulas with thousands of Greek letters. The models are run on supercomputers using millions of lines of code and are back-tested on millions more data points. In other words, the same way we predict whether it will rain tomorrow.

This has proven to be an inexact science. Even the most confident climatologists agree there is no sure way to predict the outcome of SRM. Everyone agrees, however, that SRM would be the most important and consequential action to date taken by any creature in the history of the earth. And we are betting on a weather forecast.

As mentioned, continuous spraying would be necessary since rain washes particles from the skies. If international disagreements on SRM or wars halted spraying, then global temperatures could suddenly spiral, initiating an out-of-control positive feedback loop that could potentially trigger a runaway greenhouse effect. That is why, once started, it would be necessary to spray without interruption. Even without political disruptions, our history on large, technologically complex projects suggests that a zero-failure rate is a stretch goal. Apollo 1 and 13, Challenger, Columbia, Three Mile Island, Chernobyl, and Fukushima come to mind. SRM requires 100 percent reliability forever, which is a long time.

This would be especially true if the nations of the world, comforted by this quick and easy fix, placed no further limits on greenhouse gas emissions. When seat belts were mandated in cars, the number of auto accidents increased, as drivers on average drove faster.[36] As a kind of global warming seat belt, SRM may encourage more of the behavior it was designed to counteract. An even greater concentration of greenhouse gases in the atmosphere would increase the severity of a temperature spike due to an interruption in SRM measures. Another challenge is that while SRM can reduce the amount of sunlight that falls on land and water, it would not cut the amount of greenhouse gas in the atmosphere. If greenhouse gas continues to build, the oceans over time will absorb more CO_2, which has the effect of increasing acidity.

Since preindustrial times, the acidity of the ocean has been steadily rising. The world's oceans are already 30 percent more acidic than preindustrial levels.[37] The increasing levels of acidity of the world's oceans are harmful to aquatic life. Scientists debate the pH level at which the oceans will become lifeless, but all agree that at some level, acidity is toxic to fish.[38] Even a lifeless ocean would not spell the end of human life or sushi restaurants: We could still survive on fresh water collected from rainfall and farm fish in large vats on land. However, a highly acidic ocean, devoid of all life, may have unforeseen consequences for the planet. For example, ocean life sucks up significant amounts of CO_2, one of the many complex atmospheric feedback loops that determine our planet's climate.

The National Research Council in the UK has studied geo-engineering as a way to reset the global thermostat. One of the scientists who co-authored this study concluded that a plan to geo-engineer the global climate, given the potential risks, was "wildly, utterly, howlingly barking mad."[39]

Initial Geoengineering Efforts:
The Villainous Greenfinger

Some small-scale attempts at geoengineering have already been attempted.

During the summer of 2012, Russ George, a wealthy San Francisco technologist, dumped 120 tons of iron sulphate into the Pacific Ocean. His intention was to create a plankton bloom ten thousand miles square to absorb carbon dioxide in order to reset global temperatures.[40] George had been denied a permit to dump the iron sulphate by the US EPA. So, he chartered some boats and sailed into international waters off Canada for the unauthorized releases. This attempt at geoengineering drew sharp responses from the scientific community for the potential unforeseen consequences to the climate and ocean environment. In the end, no material impact on the amount of greenhouse gases in the atmosphere or environment was detected, and his effort failed. However, George did become famous. Among environmentalists, he is known as Greenfinger, after a villain in a James Bond movie.

Besides Greenfinger, there have been other geoengineering efforts, although not on a global scale, to alter the climate. China has attempted to increase precipitation on the Tibetan plain by up to 10 billion cubic meters annually.[41] China has built thousands of small chambers in the Tibetan mountains to burn a solid fuel that emits silver iodide, a cloud-seeding agent with a crystalline structure, into the skies. Once airborne and swept into the clouds by mountain winds, the silver iodide acts to induce snowfall. The additional snowpack adds to the glaciers that melt and feed the Yellow, Yangtze, and Mekong Rivers.

The head of this effort was quoted as saying:

> *Modifying the weather in Tibet is a critical inno-*
> *vation to solve China's water shortage problem. It*

will make an important contribution not only to
China's development and world prosperity but also
the well-being of the entire human race.[42]

Because global warming poses an existential threat to some countries, we should not be surprised if a desperate nation launched SRM without the approval of others in an attempt to forestall economic and political collapse.

This will be controversial, as the optimal temperature of the global thermostat varies by country. Given the potential huge economic and social impacts, nations will be strongly motivated to prevent others from messing with their weather. In this respect, geoengineering is a new kind of WMD and could be met with threats of war or even nuclear retaliation. Nations suffering the most from global warming may use WMD to defend their geoengineering efforts, while nations suffering less may use WMD to stop them.

China, India, and Russia: Not on the Same Page

China will suffer significant environmental and economic damage if sea levels and temperatures continue to rise.[43] As sea levels increase, China will need to eventually resettle tens of millions of people from its coastal cities.[44] The Chinese government has acknowledged that the increasing severity of droughts in northern China are the result of climate change, and warmer temperatures have negatively impacted agricultural production.[45] An even more pressing issue for China is water: A billion people could be affected by a shortage of drinking water due to global warming. The Chinese government has acknowledged that the glaciers that are the primary source of water for their rivers are disappearing at a rate of 7 percent per year.[46]

In the case of China, it is hard to predict how the country's political leadership would react if the effects of global warming threaten to collapse the economy and threaten Communist Party rule. China has the resources to geoengineer the climate and would be difficult to stop, short of nuclear war. China has already demonstrated in Tibet a willingness to attempt to alter the climate, albeit on a small scale, without consulting others in the international community.

India is another country that will suffer from rising temperatures: Crop yields could drop by about one-third.[47] This will be particularly devastating, as a large portion of the Indian population are peasant farmers.[48] A 25 percent decline in crop yields would lead to an estimated 250 million people without sufficient levels of nourishment.[49]

However, the right temperature for China is too cold for India.[50] Thus, the world's two most populous nations will have differing views on how much to chill the planet. And Russia will be opposed to any reduction in heat, since it is the only major country that significantly benefits from global warming. With a sufficient rise in global temperatures, much of northern Russia becomes agriculturally productive and certainly more hospitable. Siberia may never become a popular tourist destination, but it could become the breadbasket of Asia and other parts of the world.

Hence, India, China, and Russia will have very different views on global warming. Any attempt to alter the global climate could be met with a military response by one or all of these nations.

The director of the CIA warned the US Congress about the potential military implications:

> *Northern Eurasia stability could be substantially affected (by global warming) . . . China has never*

recognized the Czarist appropriations of Chinese territory (Siberia) . . . which is rich in oil, gas and minerals. A small Russian population might have substantial difficulty preventing China from asserting control . . . the probability of conflict between two destabilized nuclear powers would seem high.[51]

The United States and Central America: Beyond the Wall

The United States will not suffer directly from global warming to the same extent as many other countries. It is true that the breadbasket of the nation, the Central Valley of California, where one-quarter of all the produce sold in the United States is grown, might have to be abandoned if rivers run dry in the summer.[52] In that case, the Northeast, Northwest, and Rocky Mountain regions would all experience gains in crop yields, partially offset by declines in agricultural production in the rest of the country.[53]

For the United States, the much greater impact from global warming will be from south of the border. With rising temperatures, Mexico and Central America could expect to experience declines in crop yields from 19 to 25 percent.[54] A meaningful increase in temperature in these countries will turn farmland from Costa Rica to southern Mexico into dust bowls.[55] If this occurs, waves of desperate immigrants, forced to flee north by starvation and political unrest, could overwhelm the southern border.

In response, the United States could seal the southern border. But then the economy and government of Mexico would collapse as starving refugees from Central America flood the country, and cross-border trade and remittances dry up. Abutting the southern border of the United States would be a failed state of more than 130 million people. The United States could open the border to relieve the pressure on Mexico, but then tens of millions of people would immediately stream into the United States from Central America and Mexico, with many more to follow.

If the United States decided there was no choice but to defend the southern border, then the US military would need to be deployed. Imagine the prospects of a hundred-foot-high concrete wall, protected by land mines and hovering helicopter gunships circling in the skies overhead, with battleships patrolling off the coasts of California and the Gulf of Mexico. Thousands of people attempting to enter the country by land and sea might have to be killed each month. But the militarization of the southern border will be the only way to stop a mass migration of refugees if global warming collapses the nations of Central America and Mexico. Faced with the prospects of the preceding scenario, the United States, Mexico, and the nations of Central America would have an incentive to geoengineer the planet.

Measures to Combat Global Warming

As discussed previously, I believe greenhouse gas emissions and global temperatures will continue to rise for the foreseeable future, given rising standards of living in the developing world and the economic, political, and intergenerational issues related to global climate agreements. Our current path leads eventually to a global climate crisis and then, as a last resort, geoengineering and all the dangers it entails.

I believe our best and perhaps only realistic alternative path is to invent our way out. If carbon emissions were dramatically cut through technological innovation, then we avoid gambling on geoengineering, which is not only risky but could also spark a global conflict.

Unfortunately, today we invest far too little in "green tech."

Total US government spending on green tech research is estimated at $7 billion per annum.[56] For comparison, US government expenditures in FY 2021 are projected to be $4.8 trillion.[57] The US allocates $38 billion each year to the National Institutes of Health (NIH) for medical research.[58] In 2019, US citizens forked out $95 billion on pet food.[59]

Some maintain we should let the more efficient private sector fund green tech research and keep the government out of it. But private investors in green tech will have different objectives than the rest of us.

Private investors seek to maximize returns, which is a different goal from reducing greenhouse gas emissions. Private investors fund technologies that today rely partly upon the ability to freely pollute the skies with greenhouse gases. That is because no major country has implemented a significant carbon tax (which is politically impractical), and subsidies are constrained by stretched government budgets (which makes them fiscally impossible). Until the cost of carbon emissions is part of a business's profit and loss statements, private investment will be skewed toward technologies that contribute to a net increase in greenhouse gas emissions. By contrast, governments can decide to fund technologies whose sole purpose is to cut greenhouse gas emissions.

In addition, private investors seek a return on their investment. But it would be better to give away clean technology, royalty-free, if the sole objective is to reduce greenhouse gases. Imagine there was a potential new technology that dramatically cut greenhouse gas residential emissions in India but few of that country's citizens could afford it. A private investor would not fund research to develop this technology since the returns could not justify the investment. But a government could, and then give the technology away.

Hence, the private sector, in my view, cannot be relied upon to invent our way out. Without a carbon tax or heavy government subsidies, private investors will do exactly what they say they will do, which is to maximize returns by funding technologies that partially exploit the government's failure to charge for polluting the atmosphere. The goals of private investors are not solely to decrease carbon emissions. And few private investors are understandably up for giving stuff away.

If we cannot count on the private sector to fund green tech, then we will have to turn to governments. In my view, the US government should take the lead and make significant investments in green tech. The sole purpose of those investments should be to cut greenhouse emissions, and we should be willing to give away anything that works. And the US government should encourage other governments around the world to do the same.

On May 25, 1961, John F. Kennedy went before a joint session of Congress and established for the nation the ambitious goal of landing an American on the moon. Today, the US still spends $22 billion a year on NASA.

The United States should set up the equivalent of NASA for climate change. Funded with an annual budget of $50 billion, the purpose of a "climate NASA" would be to invest in and develop the next generation of breakthrough technologies to reduce carbon emissions, including funding initiatives specially targeted to help developing countries, the source of most of the growth in greenhouse gases in the future.

The previous generation can brag that they invented the technologies that took Americans to the moon.

Our generation could aim even higher. Our moon shot could be to save the planet.[60]

Conclusions

Human emissions of greenhouse gases have not yet triggered a runaway greenhouse effect, despite our best efforts. But this may just be a case of survivor bias. If we had triggered a runaway greenhouse effect, there would be no humans left to lament that we finally went too far. Yet, we continue to pollute the atmosphere, the greatest commons in human history. Given the complex positive and negative feedback loops that govern the climate of our planet, we cannot know with certainty how far away we are from the tipping point that, once crossed, will lead to the extinction of our species.

Unfortunately, the challenges inherent in global climate agreements and domestic politics will severely constrain our ability to combat rising temperatures for the foreseeable future. So, we will likely continue to spew increasing amounts of CO_2 into the atmosphere as the per-person emissions of greenhouse gases rise in the developing world. Our approach to the global climate has been comparable to that of a driver in a car speeding down the road who cannot see out the windshield and decides the best thing to do is give it more gas. We might consider a tap of the brakes until the visibility on our climate future is clearer. If we keep going, we will eventually hit something.

As the planet warms, the chances that a desperate nation risks geoengineering increase. The major nuclear powers are definitely not on the same page regarding the right temperature at which to set the global thermostat. Many nations will feel this issue is worth fighting for and be willing to do so. Hence, the potential for war will increase in the future, particularly if nations such as India and China believe their choices are either to launch a geoengineering program, reset global temperatures, or watch as their people starve and their regimes collapse. The United States may feel the same with a failed state of more than one hundred million destitute, starving immigrants looking to cross an exposed southern border.

As we consider the existential risks to our species from nuclear weapons and global warming, it would be comforting if we had evidence that intelligent life had evolved on other planets, acquired nuclear weapons, burnt fossil fuels, and survived. In terms of the analogy of the Greek ships, evidence of intelligent life on other planets would be the equivalent of sailors on ship Earth returning to port and observing ships from other planets returning safely after similar voyages.

But as of mid-2021, there is no such evidence.

Given hundreds of billions of planets swirling around hundreds of billions of stars, that seems odd. There should be somebody out there. It would be encouraging if in the last 14 billion years at least a couple civilizations had sprung up on other planets, developed advanced technologies, and lived to tell the tale. It would be great news if the galaxy looked like an episode of *Star Trek* with humans, Vulcans, and Klingons, or a scene from *Star Wars* with various creatures from across the universe driving flying cars and packing ray guns.

For over half a century, fields of radio telescopes, and more recently satellites, have turned their electronic ears to the sky, listening to the stars. So far, we have heard only an eerie silence. There are no signs of intelligent life beyond our lonely planet.

So, why have we not heard from Spock, or at least ET?

CHAPTER 10

Fermi's Paradox: Where Are They?

Enrico Fermi: Technological Civilizations and ET

On a hot summer day in 1950, Enrico Fermi (1901–1954), one of the most brilliant and famous scientists of the twentieth century, was walking to lunch at Los Alamos National Laboratory in New Mexico with Edward Teller, Emil Konopinski, and Herbert York, physicists working with Fermi on the next generation of atomic weapons.[1] Fermi asked his meal companions a simple question: "Where are they?"[2] Fermi had heard news reports of UFO sightings, and his curiosity had recently been sparked by a cartoon in the *New Yorker* that showed aliens stealing New York City trash cans.

The basis of Fermi's question, known today as Fermi's paradox, was that given hundreds of billions of planets orbiting hundreds of billions of stars, there must be other intelligent life in the universe. Fermi asked the question the way he did because he believed intelligent life must have evolved on other planets, and the paradox is why we have not heard from them.

Born in Rome, Fermi studied physics at the University of Pisa and, after receiving a PhD, taught at the University of Florence and later the University of Rome. In 1929, Italian prime minister Benito Mussolini appointed Fermi to a position at the newly created Accademia d'Italia. But the anti-Semitic policies of the Italian and German regimes troubled Fermi, especially since his wife was Jewish. So, in 1938, Fermi picked up his Nobel Prize in Stockholm and boarded a ship with his wife to the United States. Upon arrival, he accepted a position at Columbia University.

After the United States entered World War II, the US government's nuclear research was consolidated under Fermi at the University of Chicago. On December 2, 1942, in a squash court under the stands of Stagg Field, Fermi achieved the first sustained nuclear reaction within the artificial reactor Pile-1. In 1944, Fermi was assigned to Los Alamos, where he played a critical role in the development of the first nuclear weapons. After the war, he continued his research at the University of Chicago and died of stomach cancer in 1954. An atomic research institute at the University of Chicago was later established in his name. (Robert Wald, son of Abraham Wald, is a physics professor at the Enrico Fermi Institute.)

Fermi was called by some of his colleagues "The Pope." This was due both to his Italian heritage and his extreme self-confidence. Fermi was famous for arriving at the answer to a complicated physics problem before his colleagues could pull a pencil from a pocket protector. He did this by a series of estimates and quick calculations in his head, which in many cases proved remarkably accurate. He once stated: "I can calculate anything in physics within a factor of 2; to get the numerical factor in front of the formula right may well take a physicist a year to calculate, but I am not interested in that."[3]

One of Fermi's most famous back-of-the-envelope estimates was that of the yield of the first atomic bomb. On July 16, 1945, at 5:29 a.m., the first nuclear weapon was detonated in the desert of New Mexico. Fermi dropped strips of paper in front of the blast wave and then measured the distance they were blown by the explosion. He made guesses about the factors that determine the distance strips of paper fall and, with some quick mental math, calculated the yield at ten kilotons of TNT. This estimate was in the ballpark of the actual yield calculated weeks later by a team of physicists and mathematicians.

The Fermi Method

This idea of getting to a "good enough" answer quickly based on a few simple calculations has become known as the Fermi method. The degree of Fermi's self-confidence can be seen in the fact that he relied upon the Fermi method for parts of his calculations that day in 1942 when the first fission reaction was triggered. At that time, it was not known whether nuclear reactions, once started, could be stopped. Most people would require more than a best guess before lighting a nuclear fuse for the first time. It is a testament to Fermi's brilliance (or survivor bias) that the first sustained nuclear reaction was not his last.

An example of the Fermi method is how to estimate the number of piano tuners in Chicago in 2009.[4] Assume Chicago at that time had a population of nine million. The following series of guesses could be made:

1. Two people per household

2. One household in twenty has a piano

3. A piano is tuned once a year

4. It takes two hours to tune a piano

5. Piano tuners work forty hours per week, fifty weeks per year

Based on one known assumption, the population of Chicago, and five more guesses, we arrive at an estimate of 225 piano tuners. In 2009, there were actually 290.

The Fermi method works surprisingly well if the errors in the assumptions are randomly wrong, which is often the case. The basic idea is the guesses that are too high will tend to cancel out those that are too low. For example, misestimates in the number of people per household and the number of households that have a piano are uncorrelated and therefore will, on average, offset each other.

So, as the physicists sat down to lunch on that hot summer day in 1950 in the cafeteria of Los Alamos National Laboratories, they began to apply the Fermi method to the question of why we have not heard from aliens. They debated the guesses that needed to be made, such as the number of galaxies in the universe, the average number of suns within a galaxy, and the number of Earth-like planets circling sun-like stars. They speculated on the odds of life beginning and then developing the intelligence required for space travel, or at least to beam electronic signals in our direction. The physicists did not come to any conclusions, other than they recalled that Fermi said we should have heard from aliens by now.

Two Possible Solutions to the Paradox

Since then, a number of solutions to Fermi's paradox have been proposed. Two stand out.

One proposed solution is the zoo hypothesis. This theory proposes that aliens are watching us from space in much the same way humans gawk at animals in a zoo. The lions, tigers, and zebras in a zoo do not realize they are entertainment, or even that they are locked up. Many of the pens look remarkably similar to their natural habitats, and the food is better. The wild animals in game parks are even more clueless. The zoo hypothesis contends that aliens are out there, but we don't realize it since we are comfortably caged in zoo Earth.

I would hope an advanced alien species would have better things to do. On the other hand, zoos are quite popular with humans, especially young children. If the real purpose of our species is to provide entertainment for alien kids, then we will have to rethink our place in the universe.

Another solution is the simulation hypothesis. This theory is that technologically advanced aliens have created a galactic simulation of a survival game, again for their amusement, in which we are the players. Similar to the zoo hypothesis, we are unaware that we are actually autonomous programs, part of some cosmic alien PlayStation.

Unlike the zoo hypothesis, the simulation hypothesis has a number of supporters within the physics community. It is also popular with several technology billionaires. The theory behind the simulation hypothesis is that for superintelligent aliens, constructing a fake universe within computers on their home planet would be a lot easier (and cheaper) than forging another universe, with galaxies, suns, planets, black holes, etc., out of the physical world. Once built by aliens, a program that was smart enough to simulate a fake universe with billions of creatures could presumably replicate itself, and the replicated fake universe could create other fake universes and so on until the world is populated with billions of fake universes. In this scenario, the ratio of fake to real universes is an exceptionally large number, and therefore it is much more likely we are living in a fake universe than the real one.

Of course, a theory that cannot be proven false is not really a theory. The opponents of the simulation hypothesis argue it is not worth talking about since there is no way for us to depart our fake universe in *Matrix* red-pill style and observe that it is in fact a fake universe. However, those who favor the simulation hypothesis beg to differ. They contend that the "glitches" in modern science, such as the conflicts between relativity and quantum mechanics (see chapter 11), are evidence we are living in a computer simulation. The glitches, they claim, are bugs in the alien program. The reason superintelligent alien programmers do not patch these glitches is that, like their human counterparts, it is easier to declare that these bugs are new product features.

But a word of caution. Those who support the simulation hypothesis may wish to avoid delving too deeply into the matter to find proof of their beliefs. Let's say the advocates of this theory were to convince the rest of us that the simulation hypothesis is true and, as a result, we all start behaving differently, or even worse, in unentertaining ways. This creates the risk that our alien creators lose interest and decide to do a hard reset on our simulation.

The Known Unknowns

If we put aside the zoo or simulation hypotheses, then resolving Fermi's paradox is challenging. Let's assume we somehow know approximately how many Earth-like planets there are circling sun-like stars.[5] We still don't know the odds of the evolution of intelligent life on one of those planets. As of today, we cannot even hazard a guess.[6]

Science has yet to determine how life began on earth. Ninety-nine percent of the weight of all living organisms comprises six ingredients: carbon, hydrogen, nitrogen, oxygen, phosphorus, and sulfur.[7] Those building blocks of life are prevalent throughout the Milky Way.[8] To spark the first replication of cells, presumably energy from the sun was required. However, how this primordial soup was excited by the sun's rays to form the first unicellular life is a mystery. Various attempts to replicate life have been made, such as those by Harold Urey and his student Stanley Miller, who independently synthesized amino acids in a lab dish.[9] But that is a long way from creating even the simple single-cell creatures that began life on earth billions of years ago.

Even once unicellular life develops, we still don't know how likely it is that intelligent life will evolve billions of years later. All we know is that hundreds of millions of species have come and gone on this planet. Yet only one species developed the technological chops to construct space shuttles and reach the cultural heights of reality TV.

Known unknowns, such as the odds intelligent life will evolve on an Earth-like planet, are an important reason why attempts to resolve Fermi's paradox have failed. The Fermi method works if we can make some reasonable guesses, such as the number of pianos per household. But if we don't know whether there are 1 or 86,000 pianos per household, then the methodology of the Fermi method breaks down. In this case, the problem is we don't know whether the odds of the evolution of intelligent life on a planet like ours are one in a million, billion, or trillion. Using the Fermi method to resolve the Fermi paradox has a nice ring to it, but it is the wrong approach to the paradox, given the degree of uncertainty of some of our assumptions.

In fact, applying the Fermi method to the Fermi paradox is another example (this will not be surprising to the reader) of survivor bias. The Fermi method is all about looking for survivors, aliens living on other planets. Rather than using the Fermi method focusing on the evidence of survivors on other planets, we should consider the evidence concerning non-survivors. But before we do, let's consider the one data point we have about survivors and what it can tell us.

The Mediocrity Principle:
One Is Better Than None

The fact that we exist proves intelligent life is possible. The question then becomes what we can infer from the fact of our existence about whether there is anybody else out there, even though we only have one data point. In statistics, the idea that some data are better than none is known as the mediocrity principle. The basic idea of the mediocrity principle is that even one or two samples drawn from a larger set are more likely than not to be representative of the larger set.

To illustrate the mediocrity principle, imagine we want to guess whether there are more green or red balls in a box, but we are not allowed to peek inside. Furthermore, we have to make our guess on the basis of drawing only one ball from the box. The mediocrity principle says that if we draw a green ball, there are likely more green balls than red balls in the box. Similarly, if we draw a red ball, there are likely more red balls. Of course, our estimate does not have a great deal of certainty given a sample set of one, and the more balls we draw, the more confident we are of our conclusions. But even one draw is better than none. So, the theory goes, a single draw is more likely than not to be representative of the total set, or "mediocre."

Now replace the balls in a box with billions of planets, some with intelligent life and others without. Our one draw came up with intelligent life—ourselves. Based on our one draw, the mediocrity principle says that more likely than not intelligent life is representative of the larger set of all planets.

You may question the validity and usefulness of the mediocrity principle. But it has been employed by mathematicians for more than a century to solve real-world problems. It has proven itself time and time again to be remarkably accurate. The most famous example is known as the German tank problem.[10]

The German Tank Problem

During WWII, the Allies wanted to know how many Nazi tanks were produced each month. However, data on wartime production of Nazi tanks was challenging to acquire, and for any given factory, "intelligence data could be found that would support any given output."[11]

A review completed after the war concluded:

> Secret sources and interrogations yielded large masses of contradictory reports. Some of these reports undoubtably were accurate, but it was impossible for the analyst to discover these in the flood of rumors and false statements.[12]

The best estimates of US intelligence were a production rate of Nazi tanks of about 1,400 per month, based on the reports of spies and captured German soldiers.[13]

But the Army Logistics Office, or ALO, in London had a different view. The mathematicians at the ALO had data on several tanks that had been captured, specifically the serial numbers of the gearboxes. The ALO knew that the Nazis produced tank gearboxes in numerical order. Based on the mediocrity principle, the ALO assumed that the serial numbers of the few tanks that had been captured were most likely to be about average, or "mediocre."

Here's an example of how the ALO projected German tank production using simple numbers for ease of calculation. The method the ALO used was to average the difference between the highest and lowest serial numbers and then add that to the highest serial number. Assume the Allies captured two Nazi tanks with gearbox serial numbers of 150 and 100. The average difference between these two numbers is 25 (150 minus 100 divided by 2). Adding 25 to 150 yields an estimate of 175 tanks produced that month.

The actual statistical formula used by ALO was slightly different and based on more complex mathematics. Nevertheless, the mathematicians relied on the basic idea of the mediocrity principle for their estimates—that a given serial number was "mediocre." Using the samples of a handful of captured German tanks, the ALO estimated that the German tank production was about 246 per month, or 80 percent less than US intelligence estimates. This lower estimate was one reason the Allies were comfortable proceeding with the D-Day invasion. After the war, German records showed the actual monthly number was 245.[14]

The ALO also used the mediocrity principle to estimate V-1 rocket production. The ALO estimated the total production of V-1 rockets to have been about thirteen thousand. The actual number was determined after the war to have been about twelve thousand.[15] Similar estimates were undertaken for the production of light, medium, and heavy trucks. An analysis after the war demonstrated the estimates were, on average, within 22 percent of the actual figures.[16]

Given the demonstrated ability of the mediocrity principle to solve real-world problems, this theory has been applied to Fermi's paradox. Those who employ the mediocrity principle to gain insights into Fermi's paradox argue that the one sample we have—ourselves—indicates that the existence of other intelligent life is more likely than not. Our one draw from the box containing planets with and without life came up with not just life, but intelligent life.

The Mediocrity Principle and Alien Volunteers

Unfortunately, we can be deceived by survivor bias when we apply the mediocrity principle to Fermi's paradox.

The mediocrity principle assumes that samples are drawn randomly. In the instance of the box with green and red balls, this is the case: The likelihood of drawing a green ball is proportional to the number of the green balls in the box, and the same goes for red balls.

But our observation that life exists on earth is not a random sample. Our sample of one suffers from survivor bias. We are both the observer and survivor, as the planet from which we are drawing is ours. On all the lifeless planets in the Milky Way, nobody is thinking about applying the mediocrity principle to resolve Fermi's paradox.

On the other hand, if an alien from another galaxy picked our planet at random from all the other planets in the universe, then the instance of intelligent life on earth would not be subject to survivor bias. However, we could not have hired the alien because that would have biased our sample to include only those galaxies that had survived and had aliens to hire. To eliminate survivor bias, the alien needs to have stepped forward independently, unprompted by *Homo sapiens* attempting to solve Fermi's paradox. Without the aid of an alien volunteer, those who argue for the existence of other intelligent life in the universe based on the mediocrity principle are failing to correct for the survivor bias. On all the sterile planets in the Milky Way, no such samples by alien volunteers are being taken.

Hence, we should not apply the mediocrity principle to Fermi's paradox. Unfortunately, there is not much we can conclude concerning Fermi's paradox based on the evidence we have from our sample of one.

But we can learn something from the non-survivors.

A Medical Test: Evidence of Non-survivors

Those seeking to resolve Fermi's paradox have traditionally sought evidence that aliens exist. In effect, they are looking for survivors. I believe an equally valid approach is to seek confirmation that aliens don't exist. That is, we should focus on the evidence of non-survivors, which could tell us just as much.

The analogy of a medical test illustrates this point.

Suppose that doctors discover a genetic mutation that allows a person to live forever. Let's say that genetic mutation is marked by a hypothetical "forever gene," which can be detected by a simple medical test. Although simple, that medical test does have some peculiar characteristics: The false positive rate is zero, but the false negative rate is unknown. In other words, if a person tests positive, it is 100 percent certain that person has the forever gene and will live forever. If a person tests negative, then the odds of possessing the forever gene are unknown.

Imagine a person takes the test, receives a negative result, and tells a friend of his disappointment. The friend's reaction is that she is sorry for him but also heartened to know that she will likely live longer than he will. This upsets him. He not only wants to live a long life, but he also believes she is wrong in her belief. How can she insensitively claim she expects to have a longer life when she has not even taken the test?

She is insensitive, but right. There is some possibility that he has the forever gene, but the probability that he has it is unknown, other than that it is greater than zero. After taking the test, he knows one thing he did not before—he tested negative. Therefore, the odds of him having the forever gene are now lower than they were before he took the test. Anyone who takes the test gets either a true positive or a negative result, which could be either true or false. (Recall there are no false positives.) So, he should realize that because he is not true positive, this lowers the probability of possessing the forever gene compared with his friend who has not tested.

Looking for evidence of intelligent life is similar to the forever gene test. If we discover intelligent life, then that is the equivalent of a positive test result. Once we meet ET, we are 100 percent certain he exists. If we don't discover intelligent life, then that is the same as a negative test result. After the test, we still don't know the odds of whether ET exists or not, but we do know the odds are lower than before. Additionally, the more times we test and get back negative outcomes, the more likely there is no ET. With multiple negative test results for alien life over half a century, the odds of other intelligent life in the universe are lower than they have ever been. And with each passing year in which we don't detect alien life, the odds that somebody is out there diminish. So far we have recorded zero positive test results and millions of negative ones.

Therefore, based on more than a half century of the equivalent of negative test results, our best guess is that aliens don't exist. There is no evidence of survivors but lots of evidence of non-survivors. The lack of evidence of survivors is not proof that there are no other forms of intelligent life in the universe. But the limited evidence we have suggests we are alone.

For sixty-six years, attempts to resolve Fermi's paradox have employed Fermi's method to look for survivors. However, we have shown Fermi's method has inherent problems due to our inability to make reasonable assumptions about the known unknowns, such as the odds intelligent life will evolve on an Earth-like planet. Other methods have focused on survivors, invoking the mediocrity principle to assert that evidence of our existence implies the existence of other intelligent life in the universe. But we can be fooled by the winners (us) when applying the mediocrity principle by failing to adjust for survivor bias. In my view, a better approach is to focus on the non-survivors and to not employ Fermi's method. I believe we can learn just as much or more from the non-survivors—the billions of observations we have of lifeless planets.

But then how do I explain hundreds of billions of potentially life-sustaining planets with no signs of intelligent life? Even if the odds of intelligent life developing on a given planet are exceptionally low, surely an intelligent alien species evolved at least once or twice? Fermi himself believed intelligent life must have arisen at some point, given the sheer overwhelming number of other worlds in our galaxy and the universe at large.

My Guess: They Are Dead

An unstated assumption in our formulation of Fermi's paradox was that intelligent life could survive indefinitely. But intelligent life could have come and gone. We may not be unique, just late to the party.

The first planets formed over 12 billion years ago. In comparison, Earth is a planetary pup, among the younger orbs in our galaxy at only 4.5 billion years of age. Other intelligent life forms may have developed hundreds of millions or even billions of years before *Homo sapiens*. And then they may have perished. Since the Milky Way is about 100,000 light years across, electronic signals from a civilization that disappeared more than 100,000 years ago would no longer stream down on Earth. And we have only been listening for about half a century. The electronic signals sent out by those earlier civilizations could have fallen on the outer membranes of bacteria or the unsuspecting ears of dinosaurs.

Therefore, I think the most plausible resolution of Fermi's paradox, given hundreds of billions of planets and the overwhelming evidence of non-survivors, is that intelligent life arose and is no more. If forced to bet, I would put my money on they are dead.

An Alternative Explanation and a Warning

Of course, another explanation is that aliens are out there, but we have been unable to connect. The universe is a big place, and a stream of electrons beaming from the far reaches of the galaxy striking our small planet is the proverbial space needle in the haystack of the cosmos. More likely, our first contact would be with self-replicating, autonomous alien probes. (This idea was first proposed by John von Neumann, which is why such vehicles are known as "von Neumann probes.") Aliens would probably send out probes before making the journey themselves, if for no other reason than to decide where to go and whether it is safe to go there. The galaxy is largely a lifeless place. After visiting a few thousand empty planets, aliens might find the next sterile planet pretty boring and decide to narrow their search to planets with intelligent life. The galaxy can also be a dangerous place, and an intelligent species would want to avoid black holes, etc. Hence, we will probably come across an alien spacecraft before we come across an alien.

But we don't know how these alien machines have been programmed, or how these autonomous, self-replicating probes might have rewired their circuits. These probes could intentionally, or unintentionally, easily wipe out life on Earth if they were sent by a civilization with technology millions of years more advanced than ours. Hence, I think we should consider whether it is a good idea to send signals into the far reaches of space, waving our electronic arms, hoping to catch the attention of a passing alien spacecraft. Given the potential risks, it seems prudent just to listen.

Alien probes may be programmed with Starfleet General Order 1, the noninterference directive. Perhaps aliens just want to swap intergalactic stories. Or maybe they are hungry.

A More Troubling Warning

As discussed previously, an intelligent alien civilization would have likely sent out probes. Even on Earth, the technology should exist in the near future to send self-replicating, autonomous probes guided by artificial intelligence to investigate other planets. If an alien civilization within the Milky Way had launched such a probe traveling at a velocity of one-tenth the speed of light, then one of

194 FOOLED by the WINNERS

these probes should have reached Earth by now. Our galaxy is 100,000 light years across and a probe traveling at 10 percent of the speed of light could traverse from one end of the Milky Way to the other in one million years.[17]

The lack of evidence of such spacecraft suggests that aliens never achieved the technological sophistication to construct von Neumann probes. That is troubling. If intelligent life was out there, then they did not survive the development of the advanced technologies that we are about to acquire.

Conclusions

In the end, I believe the simplest explanations are the best. I think there is no evidence of intelligent life because they are not out there. Given billions of earth-like planets, my best guess is they were but aren't anymore.

This was also Fermi's suspicion. As the father of nuclear energy and one of the key scientists on the Manhattan Project, he had an intimate knowledge of the destructive forces of advanced technology. According to Herbert York, during lunch that day at Los Alamos National Laboratory, Fermi said that a reason we have not heard from aliens is "a technological civilization doesn't last long enough for it to happen."[18]

Near the end of his life, Fermi echoed these same concerns, writing about the future:

> The history of science and technology has consistently taught us that scientific advances in basic understanding have sooner or later led to technical and industrial applications that have revolutionized our way of life. It seems to me improbable that this effort to get at the structure of matter should be an exception to this rule. What is less certain,

*and what we all fervently hope, is that man will
soon grow sufficiently adult to make good use of
the powers that he acquires over nature.*[19]

Fermi's concerns echo earlier chapters about the risks of advanced technologies, especially nuclear weapons and the burning of fossil fuels. To cast Fermi's paradox in terms of the analogy of Greek ships, if planets are ships and we are sailors, there are billions of ships out there but no evidence other ships returned safely. We know for certain none have sailed into our home port.

I have speculated that the most likely answer to Fermi's question posed at Los Alamos seventy-one years ago is that intelligent life in other parts of the universe has come and gone due to the acquisition of advanced technologies. If this is true, we should adjust our level of confidence about our future survival downward as our technological prowess progresses. In addition, as the evidence of non-survivors piles up year after year, we should redouble our efforts to ensure that the acquisition of advanced technologies does not threaten our survival.

Ironically, during the later years of Fermi's life, he was involved in the development of quantum mechanics, one version of which claims there is definitely intelligent life out there. In fact, a version of quantum mechanics postulates that we have not heard from these intelligent life forms because they are living and breathing in parallel universes we can neither touch nor see. If true, then we could definitely state that other intelligent life exists. But Fermi's paradox is not the question quantum mechanics has been trying unsuccessfully to answer for half a century.

It is a question about a cat.

CHAPTER 11

Quantum Mechanics: Nobody Understands

Erwin Schrödinger: More Cats

Erwin Schrödinger (1887–1961) was born in Vienna, the only child of a botanist and a mother who was half English.[1] As a consequence, he could speak perfect English, as well as German, French, and Spanish. He was an excellent student and was also able to translate Ancient Greek and Medieval German. After college, he worked as a teacher and, in 1914, was called up as a commissioned officer in the Austrian Army and served at an artillery post in Italy.

Once the war ended, he became a professor in Zurich. In 1922, he developed severe respiratory problems, which were diagnosed as tuberculosis. He retreated for nine months to a Swiss mountain village near Davos to take the "cure." At that time, high altitude and lack of oxygen was thought to stimulate the body to produce more red blood cells to fight infection. (In fact, the bacteria that causes tuberculosis depends upon oxygen to spread, so high altitude does combat the disease, although for different reasons than doctors at the time believed.) Schrödinger would return to the mountains during most summers throughout the 1920s. It was on one of these retreats, over New Year's Eve of 1926, when he conceived his greatest contribution to science.

The Schrödinger equation has been called "one of the most important achievements of the twentieth century and created a revolution in most areas of quantum mechanics."[2] The Schrödinger equation allows us to model the behavior of small particles, which was a challenge for classical mechanics. It is widely used today as originally formulated. He received a Nobel Prize for his work in 1933, and the formula is inscribed on his tombstone in Vienna.

The success of the Schrödinger equation made it possible for Schrödinger to move to Berlin in 1927 and succeed Max Planck at Friedrich Wilhelm University. However, Schrödinger disliked the Nazis and left Germany in 1933 for Oxford. When he resigned, Hitler wrote him a letter thanking him for his services to the nation. But the farewell missive also suggested that Schrödinger's departure was not unwelcome.

He arrived in Oxford with his wife and his mistress, who was pregnant with their child. After the child was born, the two women looked after the baby together. The Oxford dons were not pleased. They had been unaware of Schrödinger's living arrangements prior to offering him an academic post. They made their displeasure known. (Throughout his life, Schrödinger traveled and lived with two or more women at the same time. Schrödinger loved the company of women, and they seemed to feel the same.) Around the time the Oxford dons began voicing their disapproval of Schrödinger, Princeton University approached him, and he accepted an interview. However, when the Princeton University president learned that Schrödinger's wife, mistress, and illegitimate child would accompany him to campus, it was decided other candidates were more qualified.

So, in 1936 Schrödinger decided to return to Austria and accepted a position at the University of Graz. Years later, he described his action as "unprecedented stupidity."[3] Graz was a hotbed of Nazism and Austrians who favored unification with Germany. Furthermore, Schrödinger's departure from Germany in 1933, after winning the Nobel Prize, had not been forgotten by Hitler. Schrödinger was advised that if he wished to retain his position at the university, he needed to write a penitent letter, which he did. In the letter he called upon his fellow Austrians to embrace Hitler and his goals and stated that he had "misjudged up to the last the true will and true destiny of my country. I make this confession willingly and joyfully."[4] The Nazis published the letter widely in German and Austrian newspapers.

Soon, however, Schrödinger became uneasy about events in Germany. He was not a Nazi party member and, as a consequence, was dismissed from his post at the Prussian Academy in 1938. But his widely publicized letter foreclosed a position at an English or American university. When Schrödinger was approached about a professorship in Ireland, he arranged to travel to Dublin for an interview. But the German foreign minister, Joachim von Ribbentrop, informed Schrödinger he would not be allowed to travel to the United Kingdom.

Now doubly concerned, Schrödinger and his wife, leaving his mistress and child behind, arranged to travel to Rome under the pretense of visiting Fermi. The couple then traveled to Geneva, crossed over the border to France, and went on to Dublin. Once settled, he sent for his mistress and child, and the four of them lived in a house in Dublin for the next seventeen years. Schrödinger fathered at least two more daughters by two different mothers in Ireland and supported his extended family by lecturing at

University College. He returned to Vienna in 1956 and died of tuberculosis in 1961, aged seventy-three.

Quantum Mechanics: How Can It Be Like That?

Schrödinger is considered one of the fathers of quantum mechanics, which describes how matter and energy behave at a microscopic level. It is the counterpoint to classical mechanics, which models the macroscopic world. Not surprisingly, classical mechanics is intuitive to most, as we interact with the macroscopic world all day long. But most of us have little or no experience with the microscopic world, and hence quantum mechanics can seem quite strange.

The physicist Richard Feynman has stated:

> *I think I can safely say that nobody understands quantum mechanics . . . Do not keep saying to yourself, if you can possibly avoid it, "But how can it be like that?" because you will get down the drain, into a blind alley from which nobody has yet escaped.*[5]

Nevertheless, the predictions of quantum mechanics have proven remarkably accurate in the microscopic world. We depend upon quantum mechanics to design everything from microchips to MRIs.

Classical and quantum mechanics are fundamentally different ways in which to view the world around us. How we think about electrons is an example.

In the world of classical mechanics, an electron circles the nucleus of the atom like Earth orbits the sun. In the world of quantum mechanics, an electron is an energy wave. This energy wave is described by a formula (Schrödinger's equation) that predicts the probability of an electron's location but offers no certainty on where an electron will be at any particular time. Although not all places are possible, some places are more likely than others.

This is counterintuitive. Suppose an electron is like a ball on the top of a hill. A ball at the top of the hill exists at the top of the hill. We don't see a series of balls spread out on all sides of the hill in various probabilities of existence.

Quantum mechanics also says that electrons exist around the nucleus of an atom as energy waves at only a limited number of energy levels. Between any two of these energy levels, electrons cannot exist. Scientists know this because if electrons moved in an unbroken arc when transitioning between levels, then energy in the form of light would radiate over a continuous spectrum. It does not. Electrons "jump down" into a lower energy level by releasing energy at only specific wavelengths.

This is also counterintuitive. In the analogy of a ball at the top of a hill, the ball does not progress down a hill in a series of sudden surges. Rather, it smoothly tumbles end over end in a continuous motion as it dissipates energy.

But these are not the strangest parts of quantum mechanics.

The Copenhagen Interpretation: Moons and Spooky Actions

After discovering quantum mechanics, Schrödinger and others peered down into their high-powered microscopes at electrons, and what they saw was a bunch of particles. Several of the scientists, such as Schrödinger, had been awarded Nobel Prizes for inventing formulas that proved electrons were waves, so this required some explaining.

The explanation some of them came up with was that electrons are waves—except when we look at them. Some decided the best way to explain why the Nobels should not be returned was to say the effect of observing an energy wave "collapses" it into a particle. Until observed, an electron is an energy wave, lacking physical form. The theory that we collapse energy waves by observing them is known as the Copenhagen interpretation (CI) of quantum mechanics, which was put forth by Danish physicist Niels Bohr and others.

The CI view of the world is another way in which quantum mechanics is completely different from classical mechanics, in which the observer is independent of what is observed. In classical mechanics, a tree that falls in the woods still makes a sound, even if no one hears it. Not so according to the CI. Without the presence of an observer, the energy wave of the tree exists in a "superposition" of multiple possible states. If you believe in the CI, then the world around us is just energy waves waiting to be watched so they can become physical objects. As one physicist wrote: "We are faced with the prospect of waves that somehow magically sense that they are being observed and so decide to become particles instead."[6]

Schrödinger and his friend Einstein really did not like the CI. In particular, Einstein preferred to think visually, often working through a particularly perplexing puzzle in physics with a thought experiment involving everyday objects, and his thought experiments concerning the CI yielded contradictory outcomes.

In one instance, Einstein shared with Schrödinger a thought experiment involving two closed boxes and a single ball, "which can be found in one or the other of the two boxes when an observation is made."[7] Einstein was probing what he saw as a problem with the CI, which claimed the ball did not exist until an observer peered inside one of the boxes. Einstein argued that surely the ball was in one of the boxes before either of the boxes was opened. Along the same lines of thought, Einstein once asked a proponent of the CI whether he believed the moon existed only when he looked at it.[8]

Also, Einstein and Schrödinger did not like the idea of "non-locality," or the notion that two objects at a distance can interact with each other instantaneously. Under the CI, a wave of energy collapses into a physical object upon observation. In the case of Einstein's thought experiment, if the observer opens one of the boxes, looks inside, and sees a ball, then that causes the other box to become empty. But Einstein's theory of special relativity precluded anything from moving faster than the speed of light. If the boxes were a hundred light years apart, then Einstein thought the act of observing the box with the ball could not at the same moment empty the other box. Einstein called this "spooky action at a distance."[9]

Enter a Cat and a Box

In 1935, Schrödinger cast Einstein's thought experiment into a thought experiment of his own, which has become known as Schrödinger's cat.[10] In this variation on Einstein's original thought experiment, a cat is penned into a steel chamber along with a Geiger counter that during the course of an hour either emits a radioactive substance or emits no substance at all. The odds of the Geiger counter emitting a radioactive substance are on average 50/50, and the emissions occur randomly. When the radioactive substance is emitted, a relay releases a hammer that shatters a flask of hydrocyanic acid within the steel chamber. The cat inhales the toxic fumes and then dies. (Schrödinger's daughter years later said he did not like cats.)[11]

Schrödinger is illustrating that the CI claims the cat is neither alive nor dead until an observer opens the lid of the steel chamber to peer inside. But Schrödinger believed this was a "ridiculous case."[12] He believed that the cat was either dead or alive, killed or not killed by the hydrocyanic acid, before the observer collapsed the cat wave function by gazing into the steel box.

Einstein liked Schrödinger's version of his thought experiment.

He wrote to Schrödinger:

> *Your cat shows we are in complete agreement concerning our assessment of the character of the current theory. A function that contains the living as well as the dead cat just cannot be taken as a description of the real state of affairs.*[13]

Schrödinger himself never got comfortable with the CI.

Near the end of his life, he wrote: "It is patently absurd to let the wave function be controlled in two entirely different ways, at times by the wave function, but occasionally by direct interference of the observer, not controlled by the wave equation."[14]

Einstein expressed similar sentiments in his later years, writing that the CI "reminds me a little of a system of delusions of an exceptionally intelligent paranoiac."[15]

Over the years, others have raised more questions about the CI. One set of questions relates to what is called the measurement problem. In other words, what constitutes an observer? Does it have to be an intelligent being, such as a human? If an insect peered into the steel chamber, would the effect of an observation by the bug collapse the cat energy wave? If the insect was napping when the lid was opened, then would the cat energy wave still collapse? If a napping insect is sufficient, then what about a sleeping germ? But a germ (as far as I can tell) is not really conscious even when fully awake. Does that mean any physical object, such as the Geiger counter next to the cat, will collapse the cat energy wave and, if so, why is an observer required at all?

Many physicists today continue to adhere to the CI. While readily acknowledging the apparent contradictions pointed out by Einstein and Schrödinger, they argue that what ultimately matters is that the equations of quantum mechanics accurately model the outcomes of experiments. Hence, the CI has sometimes been labeled the "shut up and calculate" school of physics.

My own view is that most of those who adhere to the CI don't love this interpretation of quantum mechanics as much as they hate the other one.

And this is where survivor bias comes in.

The Many-Worlds Interpretation (MWI): Reach for a Gun

The many-worlds interpretation, or MWI, was formally proposed in 1957 by the physicist Hugh Everett. The MWI claims that individual wave functions do not collapse but simply branch into two or more parallel worlds. This interpretation resolves the paradoxes about the role of observers by removing observers from a central role.

In the example of Schrödinger's cat, there is a branch of the universe in which the cat lives and another one in which the cat dies. Upon opening the lid to the steel chamber and seeing a dead cat, the observer has branched off into the parallel world in which the cat dies. Under the MWI, the cat is either really alive or really dead before the observer peers into the box. The first quantum event is the emission, or lack thereof, of a radioactive substance by the Geiger counter. The second quantum event is the cat either dies, or doesn't, and the third quantum event is when the observer opens the lid. At each quantum event, the "old" world branches into two "new" parallel worlds.

Under the MWI, once the branches split, there is no more interaction between them. In the case of Schrödinger's cat, if the observer branches off into the dead cat world, there is sadly no way to go back to the alive cat world. Nor does the observer realize the alive cat world even exists, unless they are well versed in the MWI of quantum mechanics. The current version of themselves exists only in the branch of the dead cat world. There is another version of themselves who lives on in the branch of the alive cat world, but that is scant consolation. (In a related example of survivor bias, the alive cat may be inclined to believe in the MWI, and the dead cat will have no opinion.)

The MWI answers many of the objections to the CI. In the MWI version of quantum mechanics, the cat is really alive, or really dead, just in two different but equally real parallel worlds. The MWI also resolves the problem of the world not existing without an observer. Every branch of the world caused by a quantum event is independent of whether an observer is there or not. The cat lives in one parallel world and dies in the other based on the action of the Geiger counter, independent of the observer.

However, the MWI raises other issues. One implication of the MWI is the counterintuitive notion that there are an almost infinite number of parallel worlds, and different versions of you and I exist in many of them. In some number of parallel worlds, a version of you is reading this book in a slightly different way. (Or you have given up reading this chapter several pages ago in another parallel world for perfectly understandable reasons.) In any case, the proponents of the MWI argue that at least there is nothing contradictory about this version of the world, unlike the CI, in which cats are neither dead nor alive until we look at them.

Admittedly, we can't prove the other branches of the universe exist since we have already split off from them. And this is what the opponents of the MWI say is a problem with this version of quantum mechanics: A theory that cannot be falsified is not a scientific theory. Opponents argue that if the MWI cannot be proven or disproven, it is not worth talking about at all. Even some who believe in the MWI will admit there is no way to prove this particular implication of the MWI. By definition, parallel worlds that could be reached from our world are not parallel.

But not all the implications of a theory have to be subject to falsification for the theory to be believed. Imagine a theory that predicted 100 specific outcomes, and 99 of those outcomes could be proven or disproven through traditional scientific methods. Tests are conducted on the 99 outcomes, and the results of every single one of those 99 tests proves out the theory. Despite the fact that one outcome could not be tested, most would readily accept the theory as a valid scientific theory.

Both the CI and MWI make the same predictions about the outcomes of experiments in the lab. The proponents of the MWI argue that just because one prediction of the MWI theory—parallel worlds—cannot be tested does not mean we should cast the entire theory aside as unscientific. Except for the existence of parallel worlds, all the other implications of the MWI are the same as the CI and therefore equally as valid. Furthermore, the MWI is free of the contradictions that riddle the CI.

Regardless, it is safe to say that physicists, in both the CI and MWI camps, wish they had intuitive answers to the questions posed by Schrödinger's thought experiment about a cat.

Stephen Hawking, the leading physicist of the latter half of the twentieth century, reluctantly admitted while he was still alive that he was in the MWI camp. But Hawking also said: "When I hear of Schrödinger's cat, I reach for my gun."[16]

Anthropic Principles: Goldilocks and the Sailors

The MWI implies that the world in which you are reading this book is the product of a series of quantum events leading up to this moment. At each of those quantum events in the past, the "old" world branched into "new" worlds and so on to arrive at the present day. Today's world is one of the ancestors of all those other worlds, and this version of ourselves is part of that lineage. You may not realize it, but you are a branching survivor.

However, a past version of you probably did not survive every quantum event. Perhaps an unfortunate accident befell you. Suppose in the past you were strolling down a city street and unknowingly walked directly beneath a piano that was in the process of being hoisted by a crane twenty stories above. In one world, the piano slipped its straps, and in another world it didn't. Due to survivor bias, you don't realize how risky hoisting pianos can be.

Similarly, nothing in quantum mechanics nor the MWI mandates that the next universe you branch into will support human life. It could be that branching is quite a risky business, and after most quantum events we instantly perish, launched into a universe adverse to biological organisms. In fact, there are two camps of believers on the type of parallel universes that are possible: those who adhere to the strong anthropic principle (SAP) and those who support the weak anthropic principle (WAP).[17]

The adherents to the SAP claim there are a set of fundamental physical laws that constrain all that exists to "Goldilocks" universes, defined as worlds with characteristics agreeable to living organisms in general and for humans in particular. Of course, many Goldilocks universes may be lifeless, or at least devoid of intelligent beings. (Recall from Fermi's paradox we don't know the odds intelligence life will eventually evolve on Earth-like planets.) Proponents of the SAP simply claim the universe of possible universes includes only those with characteristics that are the same as ours—worlds that can potentially support intelligent life.

By contrast, the believers in the WAP say that there may be a set of fundamental physical laws that determine what universes are possible, but those laws allow for worlds quite different from our own, and Goldilocks universes are quite rare. They point to the fact that there are about twenty constants in the basic formulas of particle physics and another ten in cosmology.[18] If any of these thirty or more constants differed by even 1 percent from their known values, our universe would be unfit for life.[19] For example, a variation in the strength of the constant for gravity by one part in ten to the fortieth power would prevent the formation of stars.[20] Additionally, universes could exist with fewer than three dimensions. In a universe with one or two dimensions, complex structures, such as human brains, could not form, as the number of connections between the individual parts would be limited.[21] In a one- or two-dimensional universe, the potential for intelligent life as we know it is foreclosed. In short, the argument is that there are more ways things can go wrong than right: There are many possible states for a scrambled egg but only one that can yield a chicken.

Under the MWI, the universe is constantly evolving, continually splitting into new universes. But Goldilocks universes have a specific set of characteristics, such as three or more dimensions and dozens of physical laws that each obey a formula with constants that are "just right" for life. Those who believe in the WAP contend that Goldilocks universes are the exception—the stars literally all have to align.

Einstein stated the differences between the WAP and the SAP in terms of a Creator: "What I am really interested in is whether God could have made the world in a different way; that is, whether the necessity of logical simplicity leaves any freedom at all."[22]

Some religious philosophers picked up on Einstein's comment to claim he thought God created Goldilocks universes for us to live in. They contended that Einstein believed a divine hand fashioned our world because of the infinitesimally small odds that Goldilocks universes randomly emerge. Given an infinite number of sterile non-Goldilocks worlds and only a few that are not, Goldilocks universes must have been purposefully selected among all possible universes by some Being existing before and outside these universes. Basically, the argument is that it is more likely God exists than humans got lucky.

But this argument fails to account for survivor bias. The big bang and at least one line of branches had to be all Goldilocks universes—or we wouldn't be here to suffer headaches from thinking about quantum mechanics.

As we have seen, only a small change is required in any of the particular physical laws and features of our universe to destroy all life. And if non-Goldilocks universes occur regularly during branching, then this is not encouraging for this particular version of ourselves, survival-wise. By contrast, adherents of the SAP have a much more optimistic take on our future, as they are convinced it's Goldilocks universes as far as the eye can see.

Anything That Can Happen, Will

Fortunately, even believers in the WAP, like me, have cause for hope. One implication of the MWI is that all possible outcomes are realized in at least one parallel world. So, at every quantum event, there should be at least one jump to a Goldilocks universe, and therefore at least one version of ourselves will live on.

An analogy is to a game of Russian roulette: A gun with six chambers is loaded with one bullet, the barrel is spun, and the players take turns pointing the gun at their head and pulling the trigger. After each pull, the world branches into two new worlds: The gun fired or it didn't. After dozens of rounds, all the players will likely be dead. However, in at least one parallel universe all the players survive, as it is always possible for the gun not to fire. Under the MWI, a game of Russian roulette will continue in at least one parallel world until the players die of something else.

If the MWI is true, then different versions of ourselves live on in some number of parallel worlds, and in others we don't. In theory, there is always a version of ourselves who finds they have fortuitously landed in a Goldilocks universe. The ratio of the worlds in which we survive versus those in which we perish we cannot know for certain.

But history can be a guide.

Stanislav Petrov was assigned to bunker at Serpukhov-15 at most two nights a month. That suggests in most parallel worlds a global nuclear war began on September 7, 1983. Vasily Arkhipov was rarely assigned to Soviet submarine B-59. That would indicate in most parallel worlds our species didn't survive the Cuban Missile Crisis. Perhaps in most other universes we triggered a runaway greenhouse effect years ago.

According to the MWI, there is at least one world in which all the players in a game of Russian roulette survive. That could very well be the equivalent of the world in which we live today.

Conclusions

If the MWI of quantum mechanics is true, then the branching that occurs distorts our view of the past. In the analogy of the Greek ships observed by Diagoras, some sailors branched into the "returned safely to port" parallel world, and others did not. Those who branched into the returned-to-port parallel world are misled by survivor bias into concluding that seafaring is not a risky business. Today's world is simply the one in which we survived an almost infinite number of quantum events and subsequent branching. In other parallel worlds, we may have been targeted (again) by a space rock, defeated by the Neanderthals, or eaten by aliens. In other parallel worlds, we did not survive the acquisition of advanced technologies, such as nuclear weapons and the burning of fossil fuels.

We sometimes argue about how our world came to be or why the world is the way it is, based on religious or philosophical beliefs. If the MWI is true, then the answer is our world is just the one in which this version of ourselves survived.

In that sense, the MWI of quantum mechanics is the ultimate example of survivor bias.

Epilogue

AS WE HAVE SEEN, SURVIVOR bias often deceives us. We are misled because we focus on the winners, the successes, and the living and lose sight of those who have lost, the failures, and the dead. By failing to adjust for survivor bias, we reach the wrong conclusions.

Survivor bias impacts our daily lives. It causes us to overpay for asset managers who underperform the market and to overspend on health care that doesn't make us healthier. It convinces us to believe in ESP, that we can become the "millionaire next door," or that we can shed pounds on the latest fad diet. It persuades us that there is a simple formula to create a world-class company, that birth order determines personality, or that dropping out of college is the best way to be a billionaire.

Survivor bias warps our view of the past. It explains how the atrocities of the Nazis in Europe were heavily publicized while similar crimes committed by Unit 731 in China were covered up. It enables the justification of the unjustifiable—from the hasty decision to drop atomic bombs on Japanese cities and kill over 200,000 civilians to the slaughter of tens of millions in the former Soviet Union by Stalin and in China by Mao. It allows even democracies to silence the voices of the dead to avoid offending the ears of the living, such as American textbooks that more than a century after the end of the Civil War offer a patently untrue and openly racist account of the Old South. It explains how the horrific and painful deaths of 56 million Native Americans caused by the arrival of Europeans in the Americas are rarely spoken of and today remain unacknowledged by the US government.

Survivor bias also distorts our perspective on the evolution of *Homo sapiens*. We are here because of a series of highly improbable events, staring with a freak accident between the earth and a space rock and ending with *Homo sapiens* emerging victorious from a closely fought battle with the Neanderthals. In more recent history, we have come close to all-out nuclear war at least twice and are fortunate that we have not yet triggered a runaway greenhouse effect. In the years ahead, we face an even greater risk of extinction from the proliferation of nuclear weapons and steadily escalating carbon emissions. These threats to our future survival may explain why we have not found signs of other intelligent life in the universe. The development of advanced technologies may be a filter through which no intelligent life has yet passed.

Because survivor bias distorts our view of the past, we are overconfident in our future. We fail to recognize that our journey is but one of many paths that could have been taken. One of the lessons of the many-worlds interpretation of quantum mechanics is that our current path was probably one of the few in which we did not perish. It is the path taken because it is the only one we can observe.

If true, then we were not destined to be here; we are just the survivors. To conclude otherwise is to allow survivor bias to deceive us, like the sailors on the Greek ships observed by Diagoras who believed in the existence of the gods—until they drowned.

Most species don't make it. Ninety-nine percent of all species on earth have come and gone. But the complete history of the last humans has not yet been written. We have it within our power to be one of the exceptions—a survivor.

We have seen the deceptive influence of survivor bias in our daily lives and on our thinking. We have shown how survivor bias clouds our view of the past, which prevents us from gaining a clear view of the future.

To be a survivor, we must not be fooled by the winners.

Acknowledgments

I SHOULD STATE UP FRONT that the idea for the title came from Nassim Taleb's excellent book *Fooled by Randomness*. I find his books immensely insightful and engagingly enjoyable, including his observations on survivor bias. He and I actually worked together decades ago on Wall Street, and our conversations then were as thought-provoking as the books he has written since. It is appropriate he is quoted at the front of this book.

Jordan Ellenberg, also quoted at the front, introduced many to survivor bias and the story of Abraham Wald in his outstanding book *How Not to Be Wrong*. Jordan writes remarkably clearly about mathematics, and his explanation of Bayes' theorem should be required reading for students of any major. In addition, Gary Smith's first-rate work *Standard Deviations* offers an excellent introduction to survivor bias, and his discussion of advice books is the reason for chapter 2.

This book would not have happened without the encouragement from Gary Williams to start and then keep at it. Throughout the writing process, Gary has been a great friend and mentor as always and took on the role of reviewer on everything from content to cover. This work partly belongs to him. I should also thank Phil Vachon, who discouraged me from writing this book and thereby forced to me to take a hard look inside before committing to the project. Phil does what few are willing to do: He tells you what you don't want to hear. Jimmy Price critiqued the chapter on financial services. I argue in this book that asset managers are overpaid. Jimmy is an exception: He is really smart, highly skilled, and completely honest. In other words, worth every penny. Sanjay Saxena gave helpful insights on the chapter about medicine. Sanjay is a triple threat: an accomplished health care consultant with a brilliant medical mind who is good with people, too. Amit Singh provided a number of helpful insights into quantum mechanics and corrected several technical errors in chapter 11. His students are fortunate to have a professor who can translate physics into English.

As any writer will tell you, the completion of the first draft of a manuscript is about the halfway point on the road to holding a copy of the book in your hands. To arrive at your destination requires a team, and I am fortunate to have those at Greenleaf on my side: Justin Branch, Tyler LeBleu, and Chase Quarterman. Sally Garland was senior editor and led us throughout the entire process with immense common sense. Heather Stettler's shrewd suggestions during the development edit improved this book immeasurably. Pam Nordberg expertly copyedited the manuscript, and Killian Piraro proofread the final draft with faultless precision.

To my family, I know you often wondered, but this is what Dad was doing in the barn. Your love and support mean the world to me. This book is dedicated to you.

References

Albert Team, The. (2020). "The German Tank Problem Explained: AP Statistics Review." AP®Statistics. Last updated July 22, 2020. https://www.albert.io/blog/german-tank-problem-explained-ap-statistics-review/

Alchon, Suzanne. (2003). *A Pest in the Land: New World Epidemics in a Global Perspective*. University of New Mexico Press. Albuquerque, NM.

Allain, Rhett. (2017). "How Can a Cat Survive a High-Rise Fall?" *Wired* Magazine. August 29, 2017. https://www.wired.com/story/how-can-a-cat-survive-a-high-rise-fall-physics/

Alperovitz, Gar. (1995). *The Decision to Use the Atomic Bomb*. Alfred A. Knopf, Inc. New York.

AlphaBetaWorks. (2015). "Hedge Fund Survivor Bias." March 26, 2015. http://abwinsights.com/2015/03/26/hedge-fund-survivor-bias/

Amadeo, Kimberly. (2020). "U.S. Federal Budget Breakdown: The Budget Components and Impact on the US Economy." *The Balance*, October 29, 2020. https://www.thebalance.com/u-s-federal-budget-breakdown-3305789

Arms Control Association. (2003). "The Cuban Missile Crisis." Conversation between Georgy Kornienko and Robert McNamara. Armscontrol.org

Atkins, Robert. (1972). *Dr. Atkins Diet Revolution*. David McKay Company. New York.

Ball, Philip. (2018). *Beyond Weird*. University of Chicago Press. Chicago.

Ball, Robert. (2003). *The Fundamentals of Aircraft Combat Survivability Analysis and Design*. American Institute of Aeronautics and Astronautics. Reston, VA.

Barrow, John et al. (2009). *The Anthropic Cosmological Principle*. Oxford University Press. Oxford, UK.

Beck, Julie. (2015). "Birth Order Is Basically Meaningless." *The Atlantic*, October 21, 2015. https://www.theatlantic.com/health/archive/2015/10/birth-order-is-basically-meaningless/411577/

Becker, Adam. (2018). *What Is Real?* Hachette Books. New York.

Bertram, J. et al. (2016). "Predicting Patterns of Long-Term Adaptation and Extinction With Population Genetics." August 31, 2016. Cornell University.

Billings, Lee. (2013). "Fact or Fiction? We Can Push the Planet into a Runaway Greenhouse Apocalypse." *Scientific American*. July 2013.

Bix, Herbert. (2000). *Hirohito and the Making of Modern Japan*. HarperCollins. New York.

Bracken, Paul. (2013). *The Second Nuclear Age*. St. Martin's Press. New York.

Brand, Stewart. (2010). *Whole Earth Discipline*. Penguin Books. New York.

Brannen, Peter. (2017). *The Ends of the World*. HarperCollins. New York.

Bretscher, E. et al. (1955). "Enrico Fermi: 1901–1954." *Biographical Memoirs of Fellows of the Royal Society*. Vol 1. p. 69–78. https://royalsocietypublishing.org/doi/10.1098/rsbm.1955.0006

Brian, Denis. (1982). *The Enchanted Voyager: The Life of J.B. Rhine*. Prentice-Hall Inc. Englewood Cliffs, NJ.

Brimelow, Ben. (2018). "9 Times the World Was at the Brink of Nuclear War—And Pulled Back." April 25, 2018. *Business Insider*. https://www.businessinsider.com/when-nuclear-war-almost-happened-2018-4#october-27-1962-a-soviet-sub-almost-launches-a-nuclear-torpedo-4

Brusatte, Steve. (2018). *The Rise and Fall of the Dinosaurs*. HarperCollins. New York.

Bryan, Alex. (2016). "Performance Persistence Among U.S. Mutual Funds." January 2016. Morningstar Research. http://www.fwp.partners/wp-content/uploads/2016/09/Performance-Persistence-Morningstar-2016.pdf

Buck, Holly. (2019). *After Geoengineering*. Verso. London.

Byrne, Peter. (2010). *The Many Worlds of Hugh Everett III*. 2010. Oxford University Press. Oxford, UK.

Carlin, Dan. (2019). *The End Is Always Near*. HarperCollins. New York.

Casselman, Bill. (2016). "The Legend of Abraham Wald." American Mathematical Society. Accessed April 12, 2021. http://www.ams.org/publicoutreach/feature-column/fc-2016-06#

Chen, Stephen. (2018). "China Needs More Water." *South China Morning Post*. March 26, 2018. SCMP.com

Christian, David. (2011). *Maps of Time*. University of California Press. Berkeley, California.

Clark, Duncan. (2011). "Which Nations Are Most Responsible for Climate Change?" *The Guardian*. April 21, 2011.

Clear, James. (2020). *Best-Selling Books of All-Time*. Self-published. https://jamesclear.com/best-books/best-selling

Collins, Jim. (2001). *Good to Great*. HarperCollins. New York.

Conover, Emily. (2015). "November 1696: William Whiston's Explanation for Noah's Flood." *APS Physics*. November 2015. Vol. 24. No.10.

Coombes, Andrea. (2002). "Weight-loss Ads Too Good to Be True." September 17, 2002. *MarketWatch*. https://www.marketwatch.com/story/more-than-50-of-weight-loss-ads-deceptive-ftc-says

Crowley, A.J. (1993). "The Effect of Lorcainide on Arrhythmias." *International Journal of Cardiology*. July 1, 1993. Vol. 40. No. 2. p. 161–169.

Davies, Gavyn. (2006). "How a Statistical Formula Won the War." *The Guardian*. July 19, 2006.

Diamond, Jared. (1999). *Guns, Germs and Steel*. W.W. Norton. New York.

Diamond, Jared. (2013). *The World Until Yesterday*. Penguin Books. New York.

Diaz, Natalie. (2017). "A Native American Poet Excavates the Language of Occupation." *The New York Times*. August 4, 2017. https://www.nytimes.com/2017/08/04/books/review/whereas-layli-long-soldier.html

Ding, Iza, and Jeffrey Javed. (2019). "Why Maoism Still Resonates in China Today." *The Washington Post*. May 29, 2019. https://www.washingtonpost.com/politics/2019/05/29/why-maoism-still-resonates-china-today/

Dobbs, Michael. (2009). *One Minute to Midnight*. Vintage Books. New York.

Dormandy, Thomas. (2010). *The White Death: A History of Tuberculosis*. New York University Press. New York.

Drayton, Richard. (2005). "An Ethical Blank Cheque." *The Guardian*. May 9, 2005. https://www.theguardian.com/politics/2005/may/10/foreignpolicy.usa

Drexler, Madeline. (2002). *Secret Agents: The Menace of Emerging Infections*. Joseph Henry Press. Washington, DC.

Duffin, Erin. (2021). "U.S. Defense Outlays and Forecast as a Percentage of the GDP 2000–2030." Statista. February 8, 2021. https://www.statista.com/statistics/217581/outlays-for-defense-and-forecast-in-the-us-as-a-percentage-of-the-gdp/

Dwortzan, Mark. (2016). "How Much of a Difference Will the Paris Agreement Make?" *MIT News*. April 22, 2016. https://news.mit.edu/2016/how-much-difference-will-paris-agreement-make-0422

Dyer, Gwnne. (2011). *Climate Wars*. OneWorld Publications. Oxford, UK.

Ellenberg, Jordan. (2015). *How Not to Be Wrong: The Power of Mathematical Thinking*. Penguin Books. New York.

Elliott, Megan. (2018). "How much does the average American get paid per hour?" Showbiz CheatSheet.September 7, 2018. https://www.cheatsheet.com/money-career/how-much-does-the-average-american-get-paid-per-hour.html/

Ellsberg, Daniel. (2017). *The Doomsday Machine*. Bloomsbury Publishing. New York.

Everett, Daniel. (2017). *How Language Began: The Story of Humanity's Greatest Invention*. Liveright Books. New York.

Falconer, Bruce. (2018). "Can Anyone Stop the Man Who Will Try Just About Anything to Put an End to Climate Change?" *The Pacific Standard*. January 16, 2018.

Fazal, Tanisha et al. (2019). "War Is Not Over." *Foreign Affairs*. November/December 2019. p. 74–83.

Felton, Mark. (2012). *The Devil's Doctors*. Pen and Sword. Barnsley, UK.

Fenn, Elizabeth. (2001). *Pox Americana*. Hill and Wang. New York.

Fermi, Enrico. (2004). "The Future of Nuclear Physics." *Fermi Remembered*. University of Chicago Press. Chicago.

Ferris, Timothy. (1997). *The Whole Shebang: A State of the Universe Report*. Simon & Schuster, New York.

Finlayson, Clive. (2010). *The Humans Who Went Extinct*. Oxford University Press. Oxford, UK.

Force, James. (2002). *William Whiston*. Cambridge University Press. Cambridge, UK.

Frank, Richard. (1999). *Downfall*. Penguin Books. New York.

Friedman, Milton et al. (1998). *Two Lucky People*. University of Chicago Press. Chicago.

Gao, Chao et al. (2018). "Size, Age, and the Performance Cycle of Hedge Funds." September 2018. https://coincapp.com/pdf/Size_%20Age_%20and%20the%20Performance%20Life%20Cycle%20of%20Hedge%20Funds.pdf

Garner, Rob. (2017). "NASA Climate Modeling Suggests Venus May Have Been Habitable." NASA. August 6, 2017. https://www.nasa.gov/feature/goddard/2016/nasa-climate-modeling-suggests-venus-may-have-been-habitable

Gates, Bill. (2021). *How to Avoid a Climate Disaster*. Knopf. New York.

Gazzaniga, Michael. (2018). *The Consciousness Instinct*. Farrar, Straus and Giroux. New York.

Geoghegan, John. (2014). *Operation Storm: Japan's Top Secret Submarines*. Broadway Books. New York.

Gold, Hal. (1997). *Unit 731 Testimony*. Tuttle Publishing. Periplus Editions (HK) Ltd. North Clarendon, VT.

Goodell, Jeff. (2010). *How to Cool the Planet*. Houghton. New York.

Graff, Garrett. (2018). *Raven Rock*. Simon & Schuster. New York.

Grattan-Guinness, I. (2003). "Joseph Fourier, 1768–1830: A Survey of His Life and Work." The MIT Press. Cambridge, MA.

Grauschopf, Sandra. (2020). "Which States Have the Biggest Lottery Payouts?" *The Balance Everyday*. October 28, 2020. https://www.thebalanceeveryday.com/which-states-have-the-biggest-lottery-payouts-4684743

Greenwood, Robin et al. (2013). "The Growth of Finance." *The Journal of Economic Perspectives*. Spring 2013. Vol. 27. No. 2. p. 3–28.

Gribbin, John. (2013). *Erwin Schrödinger and the Quantum Equation*. John Wiley & Sons. Hoboken, NJ.

Grinspoon, David. (2016). *Earth in Human Hands*. Grand Central Publishing. New York.

Grote, Kent et al. (2011). "The Economics of Lotteries: A Survey of the Literature." College of the Holy Cross. Worcester, MA. http://college.holycross.edu/RePEc/hcx/Grote-Matheson_LiteratureReview.pdf

Gudzune, Kimberly et al. (2015). "Efficacy of Commercial Weight Loss Programs: An Updated Systematic Review." *Annals of Internal Medicine*. April 7, 2015. Vol. 162. No. 7. p. 501–512.

Hansen, Randall. (2008). *Fire and Fury: The Allied Bombing of Germany*. Penguin Books. London.

Harari, Yuval. (2015). *Sapiens*. HarperCollins. New York.

Harris, Errol. (1991). *Cosmos and Anthropos*. Humanities Press International. London.

Harris, Sheldon. (2002). *Factories of Death*. Routledge. New York.

Hatfield, Ken. (2003). *Heartland Heroes: Remembering World War II*. University of Missouri Press. Columbia, MO.

Hawks, John et al. (2000). "Population Bottlenecks and Pleistocene Human Evolution." *Molecular Biology and Evolution*. January 1, 2000. Vol. 17. No. 1.

Hecht, Jennifer. (2003). *Doubt: A History*. Harper. San Francisco.

Hedges, Chris. (2003). "What Every Person Should Know About War." *The New York Times*. July 6, 2003. https://www.nytimes.com/2003/07/06/books/chapters/what-every-person-should-know-about-war.html

Hemphill, William et al. (1963). *Cavalier Commonwealth*. McGraw-Hill, Inc. New York.

Hill, Napoleon. (2007). *Think and Grow Rich*. Wilder Publications. Floyd, VA.

Horn, Stacy. (2009). *Unbelievable*. HarperCollins. New York.

Horton, Alex. (2018). "How Mega Millions Changed the Odds to Create a Record-Breaking $1.6 Billion Jackpot." *The Washington Post*. October 21, 2018. https://www.washingtonpost.com/business/2018/10/17/mega-millions-tweaked-odds-create-monster-jackpots-it-worked/

Houghton, John. (2002). *Global Warming*. Cambridge University Press. Cambridge, UK.

Howes, Matthew. (2017). "Publication Bias and Clinical Trial Outcome Reporting." CenterWatch. June 12, 2017. https://www.centerwatch.com/articles/13512

Huczko, Maciej. (2014). "Alternatives to Dropping the A-Bomb in Bringing the War with Japan to an End." *Warsaw School of Economics*. Vol. 1. No. 39. p. 128–137.

Illustrated War News. (1915). "Steel-Helmeted and 'Teddy-Bear'–Coated British Officers: Ready for the Germans and for Winter." *Illustrated War News*. November 15, 1915. https://www.britishnewspaperarchive.co.uk/viewer/BL/0001862/19151117/007/0009

Ioannidis, John. (2008). "Effectiveness of Antidepressants: An Evidence Myth Constructed from a Thousand Randomized Trials?" *Philosophy, Ethics and Humanities in Medicine*. May 27, 2008. Vol. 3. No. 14. https://www.ncbi.nlm.nih.gov/pmc/articles/PMC2412901/

Ioannou, Andri. (2009). "Publication Bias: A Threat to the Objective Report of Research Results." https://files.eric.ed.gov/fulltext/ ED504425.pdf

Johnson, Ian. (2016). "China's Memory Manipulators." *The Guardian*. June 8, 2016. https://www.theguardian.com/world/2016/jun/08/ chinas-memory-manipulators

Jones, Eric. (1985). "Where Is Everybody? An Account of Fermi's Question." Los Alamos National Laboratory. UC-34B. March 1985.

Jones, Nate. (2016). *Able Archer 83*. The New Press. New York.

Joober, Ridha. (2012). "Publication Bias: What Are the Challenges and Can They Be Overcome?" *Journal of Psychiatry and Neuroscience*. May 2012. Vol. 37 No. 3. p. 149–152.

Keeley, Lawrence. (1996). *War before Civilization*. Oxford University Press. Oxford, UK.

Khan, Shamshad A. (2009). "Japan: CBW." *CBW Magazine*. October–December 2009. Vol. 3. No. 1. https://idsa.in/cbwmagazine/ Japan-CBW_skhan_1009

Kim, Sung et al. (2017). *North Korea and Nuclear Weapons*. Georgetown University Press. Washington, DC.

King, Cody. (2020). "APPA: Americans Spent $95.7 Billion on Their Pets in 2019." KSAT.com. February 28, 2020. https://www.ksat.com/news/ local/2020/02/28/appa-americans-spent-957-billion-on-their-pets-in-2019/

Kinzer, Stephen. (2019). *Poisoner in Chief: Sidney Gottlieb and the CIA Search for Mind Control*. Henry Holt. New York.

Koch, Alexander. (2019). "Earth System Impacts of the European Arrival and Great Dying in the Americas after 1492." *Quaternary Science Reviews*. Vol. 207. https://www.researchgate.net/publication/ 330825763_Earth_system_impacts_of_the_European_arrival_and_Great_ Dying_in_the_Americas_after_1492

Lane, A. et al. (2016). "Is There Publication Bias in Behavior Intranasal Oxytocin Research on Humans?" *Journal of Neuroendocrinology*. March 14, 2016. Vol. 28. No. 4. https://onlinelibrary.wiley.com/doi/abs/10.1111/jne.12384

Laquerre, Paul-Yanic. (2013). *Showa: Chronicles of a Fallen God.* Self-published.

LaRosa, John. (2019). "Top 9 Things to Know about the Weight Loss Industry." MarketResearch.com. March 6, 2019. https://blog.marketresearch.com/u.s.-weight-loss-industry-grows-to-72-billion

LeBlanc, Steven. (2003). *Constant Battles: Why We Fight.* St. Martin's Press. New York.

Lee, Sang-Hee. (2015). *Close Encounters with Humankind.* W.W. Norton. New York.

Lehrer, Jonah. (2011). "The Psychology of Lotteries." *Wired*. February 3, 2011. https://www.wired.com/2011/02/the-psychology-of-lotteries/

Levitin, Daniel. (2014). "How to Solve Google's Crazy Open-Ended Interview Questions." *Wired*. August 22, 2014. https://www.wired.com/2014/08/how-to-solve-crazy-open-ended-google-interview-questions/

Li, Eric, and Lijia Zhang. (2013). "Debunking the Myths of Mao Zedong." *South China Morning Post*. December 26, 2013. https://www.scmp.com/comment/insight-opinion/article/1390108/debunking-myths-mao-zedong

Lloyd, Marion. (2002). "Soviets Close to Using A-Bomb in 1962 Crisis." *Boston Globe*. October 13, 2002.

Lowenstein, Roger. (2011). *When Genius Failed.* Random House. New York.

Lukacs, Martin. (2012). "World's Biggest Geoengineering Experiment Violates UN Rules." *The Guardian*. October 15, 2012. https://www.theguardian.com/environment/2012/oct/15/pacific-iron-fertilisation-geoengineering

Macrae, Norman. (1992). *John von Neumann.* Pantheon Books. New York.

Malkiel, Burton et al. (2005). "Hedge Funds: Risk and Return." *Financial Analysts Journal*. Vol. 61. No. 6. p. 80–88. https://papers.ssrn.com/sol3/papers.cfm?abstract_id=872868

Mangel, Marc et al. (1984). "Abraham Wald's Work on Aircraft Survivability." *Journal of American Statistical Association*. June 1984. Vol. 79. No. 386.

Mann, Geoff et al. (2018). *Climate Leviathan*. Verso. London.

McGauran, Natalie et al. (2010). "Reporting Bias in Medical Research." *Trials*. April 13, 2010. Vol. 11. No. 37. https://link.springer.com/article/10.1186/1745-6215-11-37

McNeill, William. (1998). *Plagues and Peoples*. Anchor Books. New York.

Money, Nicholas. (2019). *The Selfish Ape*. Reaktion Books. London.

Morgenstern, Oskar. (1951). "Abraham Wald, 1902–1950." *Econometrica*. October 1951. Vol. 19. No. 4. p. 361–367.

Morton, Oliver. (2016). *The Planet Remade*. Princeton University Press. Princeton, NJ.

Moscow Times. (2019). "Stalin's Approval Rating Among Russians Hits Record High – Poll." *The Moscow Times*. April 16, 2019. https://www.themoscowtimes.com/2019/04/16/stalins-approval-rating-among-russians-hits-record-high-poll-a65245

Moyer, Liz. (2018). "Four Hedge Fund Managers Top $1 Billion in Pay as the Industry Rebounds." CNBC. May 30, 2018. https://www.cnbc.com/2018/05/30/four-hedge-fund-managers-top-1-billion-in-pay.html

Mullainathan, Sendhil. (2015). "Why a Harvard Professor Has Mixed Feelings When Students Take Jobs in Finance." *The New York Times*. April 10, 2015.

National Institutes of Health. (2015). "About the NIH." July 7, 2015. https://www.nih.gov/about-nih/what-we-do/nih-almanac/about-nih

National Park Service. (2017). "Harry S Truman's Decision to Use the Atomic Bomb." Accessed April 11, 2021. https://www.nps.gov/articles/trumanatomicbomb.htm

Neumann, John von. (1955). "Can We Survive Technology?" *Fortune.* June 1955.

New World Encyclopedia contributors. (2021). "Erwin Schrödinger." New World *Encyclopedia.* Accessed February 15, 2021. https://www.newworldencyclopedia.org/p/index.php?title=Erwin_Schr%C3%B6dinger&oldid=1006346

New World Encyclopedia contributors. (2021). "Joseph Fourier." New World *Encyclopedia.* Accessed April 11, 2021. https://www.newworldencyclopedia.org/entry/Joseph_Fourier

Nickel, Bradley. (2018). "What We Can Learn from How the Top Weight Loss Advertisers Manage Their Campaigns." Adbeat.com. Accessed April 11, 2021. https://blog.adbeat.com/what-we-can-learn-from-how-the-top-weight-loss-advertisers-manage-their-campaigns/

O'Mahony, Proinsias. (2018). "Lottery Winners and Stock Returns: The Problem of Survivor Bias." *The Irish Times.* June 5, 2018. https://www.irishtimes.com/business/personal-finance/lottery-winners-and-stock-returns-the-problem-of-survivor-bias-1.3513744

Panda, Ankit. (2018). "No First Use and Nuclear Weapons." Council on Foreign Relations. CFR.org. July 17, 2018.

Papagianni, Dimitra. (2015). *The Neanderthals Rediscovered.* Thames & Hudson. London.

Pappas, Stephanie. (2015). "Oxytocin: Facts about the Cuddle Hormone." LiveScience.com. June 4, 2015. https://www.livescience.com/42198-what-is-oxytocin.html

Parsa, H.G. et al. (2005). "Why Restaurants Fail." *Cornell Hospitality Quarterly.* August 1, 2005. Vol. 46. No. 3. p. 304–322.

Pawlowicz, Rachel, and Walter E. Grunden. (2015). "Teaching Atrocities: The Holocaust and Unit 731 in Secondary School Curriculum." The *History Teacher.* February 2015. Vol. 48. No. 2.

PBS.org. (2021). "The Story of . . . Smallpox—and other Deadly Eurasian Germs." *Guns Germs & Steel.* PBS.org. Accessed April 7, 2021. https://www.pbs.org/gunsgermssteel/variables/smallpox.html

Pearl, Judea. (2018). *The Book of Why.* Basic Books. New York.

Perry, William. (2015). *My Journey to the Nuclear Brink*. Stanford University Press. Redwood City, CA.

Peters, Tom. (2001). "Tom Peters's True Confessions." *Fast Company*. November 30, 2001. https://www.fastcompany.com/44077/tom-peterss-true-confessions

Peters, Tom (2005). "*In Search of Excellence* at (Almost) 25." *Tom Peters Blog*. https://tompeters.com/2006/11/in-search-of-excellence-at-almost-25-and-standing-tall/

Peters, Tom et al. (1982). *In Search of Excellence*. Harper & Row. New York.

Phelan, Matthew. (2019). "The History of 'History Is Written by the Victors'." *Slate*. November 26, 2019. https://slate.com/culture/2019/11/history-is-written-by-the-victors-quote-origin.html

Pinker, Steven. (2019). *Enlightenment Now*. Penguin Books. New York.

Piper, Kelsey. (2019). "This Economics Journal Only Publishes Results That Are No Big Deal." *Vox*. May 17, 2019. https://www.vox.com/future-perfect/2019/5/17/18624812/publication-bias-economics-journal

Poundstone, William. (1992). *Prisoner's Dilemma*. Anchor Books. New York.

Preston, Christopher. (2016). *Climate Justice and Geoengineering*. Rowman & Littlefield. London.

Providentia. (2019). "The Life and Times of William Whiston." Providentia. March 1, 2019. https://drvitelli.typepad.com/providentia/religion/page/3/

Pua, Derek. (2019). *Unit 731: The Forgotten Asian Auschwitz*. Pacific Atrocities Education. San Bernardino, CA.

Qazi, Junaid. (2019). "Survivorship Bias—A Danger Zone (Think Again, Do You Really Want to Miss What Is Missing!)." *Medium*. August 17, 2019. https://junaidsqazi.medium.com/survivorship-bias-a-danger-zone-think-again-do-you-really-want-to-miss-what-is-missing-de32a1d06094/

Quinn, Julian. (2020). "4.5 Survivorship bias." In: *Guides to Undertaking Research*. Version 2.1. Royal North Shore Hospital SERT Institute. November 2020. https://surgery.rnsh.org/research-guide/pdfs/4.5%20 Survivorship%20bias-%20Research%20Guide%20f2%20v101120.pdf

Ramenofksy, Ann. (1987). *Vectors of Death*. University of New Mexico Press. Albuquerque, NM.

Ramseur, David. (2017). "When an Off-Course U-2 Spy Plane out of Alaska Nearly Triggered War." *Anchorage Daily News*. June 9, 2017.

Reagan, Ronald. (1983). "Evil Empire Speech" given at the National Association of Evangelicals conference. March 8, 1983. Voices of Democracy website. https://voicesofdemocracy.umd.edu/ reagan-evil-empire-speech-text/

Reagan, Ronald. (2011). *An American Life: The Autobiography*. Simon & Schuster. Kindle version.

Reich, David. (2018). *Who We Are and How We Got Here*. Pantheon Books. New York.

Rhine, J. B. (1934). *Extra-Sensory Perception*. Bruce Humphries. Boston.

Rhine, J. B. (1957). *Parapsychology*. Charles C. Thomas. Springfield, IL.

Risi, Stephan. (2019). "Manufacturing Doubt." *Tobacco Analytics*. https://dual-momentum.com/case/ctr

Ro, Sam. (2014). "The Past Performance of a Mutual Fund Is Not an Indicator of Future Outcomes." *Business Insider*. July 13, 2014. https:// www.businessinsider.com/mutual-fund-performance-persistence-2014-7

Rohlder, Martin et al. (2011). "Survivorship Bias and Mutual Fund Performance." *Review of Finance*. April 2, 2011. Vol. 15. No. 2. p. 441–474.

Rohrer, J. M. et al. (2015). "Examining the Effects of Birth Order on Personality." *Proceedings of the National Academy of Sciences of the United States of America*. November 17, 2015. Vol. 12. No. 46. p. 14224–14229.

Rosenbaum, Ron. (2011). *How the End Begins*. Simon & Schuster. New York.

Ruggles, Richard et al. (1947). "An Empirical Approach to Economic Intelligence in World War II." *Journal of American Statistical Association.* March 1947. Vol. 42. p. 72–91.

Rummel, R.J. (1997). *Death by Government.* Routledge. New York.

Saint Michael's Hospital. (2010). "Severely Injured Should Go Directly to Trauma Center, New Research Shows." *ScienceDaily*, November 2, 2010. https://www.sciencedaily.com/releases/2010/11/101102130959.htm

Scanes, Colin. (2018). *Animals and Human Society.* Academic Press, Elsevier. London.

Scharre, Paul. (2018). *Army of One.* W.W. Norton & Company. New York.

Scheidel, Walter. (2017). *The Great Leveler.* Princeton University Press. Princeton, NJ.

Schwartz, David. (2017). *The Last Man Who Knew Everything: The Life and Times of Enrico Fermi.* Basic Books. New York.

Schwed, Fred. (1940). *Where Are the Customers' Yachts? or A Good Hard Look at Wall Street.* Simon & Schuster. New York.

Segre, Gino et al. (2017). *The Pope of Physics: Enrico Fermi and the Birth of the Atomic Age.* Picador. New York.

Shadrake, Dan. (2014). "WWI: Combat Helmet Technology—the Brodie Steel Helmet." Engineering and Technology website. June 16, 2014. https://eandt.theiet.org/content/articles/2014/06/ ww1-combat-helmet-technology-the-brodie-steel-helmet/

Sherbakova, Irina. (2019). "Vladimir Putin's Russia Is Rehabilitating Stalin. We Must Not Let That Happen." *The Guardian.* July 10, 2019. https://www.theguardian.com/commentisfree/2019/jul/10/ vladimir-putin-russia-rehabilitating-stalin-soviet-past

Shipman, Pat. (2015). *The Invaders: How Humans and Their Dogs Drove Neanderthals to Extinction.* Belknap Press. Cambridge, MA.

Shu, Pian. (2016). "Innovating in Science and Engineering or 'Cashing in' on Wall Street." November 2016. Working Paper 16-067. Harvard Business School.

S.J.Res. 14. (2009–2010). A joint resolution to acknowledge a long history of official depredations and ill-conceived policies by the Federal Government regarding Indian tribes and offer an apology to all Native Peoples on behalf of the United States. 111th Congress. https://www.congress.gov/bill/111th-congress/senate-joint-resolution/14/text

Smil, Vaclav. (2018). *Energy and Civilization: A History*. MIT Press. Cambridge, MA.

Smith, Gary. (2015). *Standard Deviations*. The Overlook Press. New York.

Stanley, Thomas et al. (1996). *The Millionaire Next Door*. Pocket Books. New York.

Steel, Duncan. (1995). *Rogue Asteroids and Doomsday Comets*. John Wiley & Sons. New York.

Sterling, Theodor. (1959). "Publication Decisions and Their Possible Effects on Inferences Drawn from Tests of Significance." *Journal of the American Statistical Association*. March 1959. Vol. 54. No. 285. p. 30–34.

Sterling, Theodor. (1965). "Statement of Dr. Theodor Sterling before the House Committee on Interstate and Foreign Commerce on Bills Relating to Cigarette Labeling and Advertising." In the American Tobacco Records, University of California–San Francisco Library. https://www.industrydocuments.ucsf.edu/docs/hncn0137

Sterling, Theodor. (1978). "Does Smoking Kill Workers or Working Kill Smokers?" *International Journal of Health Services*. July 1, 1978. Vol. 8. No. 3. https://journals.sagepub.com/doi/abs/10.2190/836R-HD65-G8JC-2CLF?journalCode=joha

Sterns, Beverly et al. (2000). *Watching from the Edge of Extinction*. Yale University Press. New Haven, CT.

Stringer, Chris. (2012). *Lone Survivors*. Times Books. New York.

Sulloway, Frank. (1997). *Born to Rebel*. Vintage. New York.

Swedroe, Larry. (2017). "How Survivorship Bias Happens: Adjusted Fund Rankings." ETF.com. February 3, 2017. https://www.etf.com/sections/index-investor-corner/swedroe-how-survivorship-biases-results?nopaging=1

Swensen, David. (2005). *Unconventional Success*. Free Press. New York.

Swift, Jonathan. (1898). "A True and Faithful Narrative of What Passed in London, During the General Consternation of All Ranks and Degrees of Mankind; on Tuesday, Wednesday, Thursday, and Friday Last." In *The Prose Works of Jonathan Swift, D.D., Volume IV: Swift's Writings on Religion and the Church, Volume II.* Appendix IV. Edited by Temple Scott. Project Gutenberg Ebook, last updated January 21, 2019. George Bell and Sons. London. https://www.gutenberg.org/files/12746/12746-h/12746-h.htm

Sy, Ramond et al. (2009). "Survivor Treatment Selection Bias and Outcome Research." *Circulation: Cardiovascular Quality and Outcomes.* September 2009. Vol. 2. No. 5. p. 469–474. https://www.ahajournals.org/doi/full/10.1161/circoutcomes.109.857938

Szanton, Andrew. (2003). *The Recollections of Eugene P. Wigner.* Basic Books. New York.

Taleb, Nassim. (2001). *Fooled by Randomness.* Texere LLC. New York.

Thomas, Andrew. (2017). *Hidden in Plain Sight 7.* Self-published.

Tillman, Barrett. (2014). *Forgotten Fifteenth.* Regnery History. Washington, DC.

Tollefson, Jeff. (2018). "Sucking Carbon Dioxide from the Air Is Cheaper Than Scientists Thought." *Nature.* June 7, 2018.

Union of Concerned Scientists (2015). "Close Calls with Nuclear Weapons." *Union of Concerned Scientists Fact Sheet.* April 2015. https://www.ucsusa.org/sites/default/files/attach/2015/04/Close%2520Calls%2520with%2520Nuclear%2520Weapons.pdf

Wald, Abraham. (1943). "A Method for Estimating Plane Vulnerability Based on Damage of Survivors." Reprint: July 1980. Center for Naval Analyses. Operations Evaluation Group. Alexandria, VA.

Wald, Abraham. (1945). "The Estimation of Vulnerability of Aircraft from Damage to Survivors." Applied Mathematics Panel. National Defense Research Committee. May 1945. AMP Report 76.1 R. SRG 448.

Wallis, Allen. (1980). "The Statistical Research Group, 1942–1945." *Journal of the American Statistical Association.* June 1980. Vol. 75. No. 370. p. 320–330.

Walsh, Bryan. (2019). *End Times*. Hachette Books. New York.

Ward, Peter. (2010). *The Flooded Earth*. Basic Books. New York.

Weissman, Jordan. (2013). "How Wall Street Devoured Corporate America." *The Atlantic*. March 5, 2013. https://www.theatlantic.com/business/archive/2013/03/how-wall-street-devoured-corporate-america/273732/

Welzer, Harald. (2012). *Climate Wars: Why People Will Be Killed in the 21st Century*. Polity Press. Cambridge, UK.

Whitney, W. et al. (1987). "High-Rise Syndrome in Cats." *Journal of the American Veterinary Medical Association*. December 1, 1987. Vol. 191. No. 11. p. 1399–1403.

Whyte, Amy. (2018). "Hedge Fund Paychecks, Revealed." *Institutional Investor*. November 8, 2018. https://www.institutionalinvestor.com/article/b1bm09m9p5mf1g/Hedge-Fund-Paychecks-Revealed

Williams, Kidada E. (2012). *They Left Great Marks on Me: African American Testimonies of Racial Violence from Emancipation to World War I*. New York University Press. New York.

Wilson, Kevin. (2007). *Men of Air: The Doomed Youth of Bomber Command*. Orion Books. London.

Winiarczyk, Marek. (2016). *Diagoras of Melos*. Walter de Gruyter Gmbh. Berlin. (English translation).

Wolfowitz, J. (1952). "Abraham Wald, 1902–1950." *Annals of Mathematical Statistics*. March 1952. Vol. 23. No. 1. p. 1–13.

Xu, Eleanor et al. (2010). "Hedge Fund Attrition, Survivorship Bias and Performance: Perspectives from the Global Financial Crisis." February 2, 2010. https://papers.ssrn.com/sol3/papers.cfm?abstract_id=1572116

Yiwei, Zhang. (2013). "85% Say Mao's Merits Outweigh His Faults: Poll." *Global Times*. December 24, 2013. https://www.globaltimes.cn/content/834000.shtml

Yost, Chad et al. (2018). "Subdecadal Phytolith and Charcoal Records from ~74ka Toba Supereruption." *Journal of Human Evolution*. March 2018. Vol. 116. p. 75–94

Notes

Introduction

1. Background on Diagoras is from the definitive biography by Winiarczyk (2016).

2. Hecht (2003), p. 9.

3. Qazi (2019).

4. To be confident in our inferences, we should count survivors and non-survivors and the incidence of prayer. For the surviving ships, we should ask the sailors of the surviving ships whether they prayed to the gods when beset by violent storms. For the non-surviving ships, we could assume prayers on every ship, prayers on none, or something in between. Unfortunately, we will never know for certain whether the sailors on the non-surviving ships prayed. We could use the portion of surviving ships on which the sailors prayed as a proxy for the non-surviving ships. Diagoras's friend would have a better argument if the non-surviving ships did not return because the sailors on those ships did not believe in the gods and therefore did not pray. But neither Diagoras nor his friend can determine with certainty the incidence of prayers on the lost boats.

Chapter 1

1. Schwed (1940).

2. Schwed (1940), p. xxxvi.

3. Schwed (1940), p. 40.

4. Weissman (2013).

5. Greenwood (2013), p. 8.

6. Whyte (2018).

7. Whyte (2018).

8. Moyer (2018).

9. Elliot (2018).

10. Schwed (1940), p. 8.

11. Lowenstein (2011) is the source for the story of John Meriwether.

12. Swensen (2005), p. 127.

13. Swensen (2005), p. 127.

14. Swensen (2005), p. 127.

15. Gao (2018).

16. AlphaBetaWorks (2015).

17. Xu (2010), p. 33.

18. AlphaBetaWorks (2015).

19. Malkiel (2005).

20. Swedroe (2017).

21. Swensen (2005), p. 203.

22. Rohlder (2011).

23. O'Mahony (2018).

24. O'Mahony (2018).

25. (0.5) to the fifth power times 10,000.

26. Ro (2014).

27. Bryan (2016).

28. In fact, a portfolio of stocks in the current DJIA will have outperformed the index itself over virtually any period in the past. That is because the deletions usually underperform the other stocks in the index.

29. Schwed (1940), p. 29.

30. Schwed (1940), p. 170.

31. Mullainathan (2015).

32. Shu (2016).

Chapter 2

1. Peters (1982).

2. Peters (2001).

3. Peters (2001).

4. Peters (2005).

5. The following analysis is from Smith (2015), p. 38–41.

6. Smith (2015), p. 39.

7. Smith (2015), p. 39.

8. Hill (2007).

9. Clear (2020).

10. Clear (2020). This list excludes the Bible, Quran, Quotations from Chairman Mao Tse-Tung, and similar works.

11. Hill (2007), p. 4

12. Hill (2007), p. 5.

13. Hill (2007), p. 114.

14. In Hill's case, other biases included his sexist assumptions that excluded all women and some men—those who did not conform to his preconceived ideas about sexuality—from the group of people he considered capable of achieving great business success.

15. Hill (2007), p. 107.

16. Stanley (1996), p. 1.

17. Stanley (1996), p. 250 describes the survey method.

18. Stanley (1996), p. 4.

19. Stanley (1996), p. 3–4.

20. Of course, the researchers, like the other business book authors, are also failing to address another weakness, that correlation is not causation. But that discussion is outside the scope of this book.

21. Sulloway (1997).

22. Beck (2015).

23. Rohrer (2015).

Chapter 3

1. Background from Brian (1982).

2. Brian (1982), p. 21.

3. Rhine (1934), p. 218.

4. Rhine (1957), p. 191.

5. Rhine (1957), p. 191.

6. Horn (2009), p. 183.

7. Horton (2018).

8. Grote (2011).

9. Lehrer (2011).

10. Grauschopf (2020).

11. Grote (2011).

12. Parsa (2005), p. 310.

13. Parsa (2005), p. 310.

Chapter 4

1. This subtitle is from a book written by one of the funniest authors ever, Christopher Buckley.

2. The Sterling Prize and background information on Sterling is from the Sterling Prize website: https://www.sfu.ca/sterlingprize/about.html

3.   Sterling (1959), p. 30.

4.   Ioannou (2009), p. 3.

5.   Sterling (1959), p. 32.

6.   Ioannou (2009), p. 3.

7.   Howes (2017).

8.   Howes (2017).

9.   Joober (2012).

10. Howes (2017).

11. Ioannou (2009).

12. Risi (2019).

13. Risi (2019).

14. Risi (2019).

15. Risi (2019).

16. Risi (2019).

17. Risi (2019).

18. Risi (2019).

19. Sterling (1978).

20. Sterling (1978), p. 2.

21. Sterling (1978), p. 2.

22. Sterling (1965).

23. Sterling (1965).

24. Ioannidis (2008).

25. Ioannidis (2008).

26. Ioannidis (2008).

27. Ioannidis (2008).

28. Ioannidis (2008).

29. Ioannidis (2008).

30. Ioannidis (2008).

31. McGauran (2010).

32. Crowley (1993).

33. Lane (2016).

34. Pappas (2015).

35. Lane (2016).

36. Lane (2016).

37. Sy (2009).

38. Sy (2009).

39. Saint Michael's Hospital (2010).

40. Saint Michael's Hospital (2010).

41. Piper (2019).

42. Piper (2019).

43. Nickel (2018).

44. Nickel (2018).

45. LaRosa (2019).

46. Coombes (2002).

47. Coombes (2002).

48. Atkins (1972), p. 57.

49. Atkins (1972), p. 56.

50. Atkins (1972), p. 17.

51. Atkins (1972), p. 18.

52. Gudzune (2015).

53. Gudzune (2015).

54. Whitney (1987).

55. Allain (2017).

Chapter 5

1. Background on Wald is from Casselman (2016), Wallis (1980), and Wolfowitz (1952).

2. Wolfowitz (1952), p. 3.

3. Morgenstern (1951), p. 365.

4. Morgenstern (1951), p. 366.

5. Morgenstern (1951), p. 366.

6. Ellenberg (2015), p. 5.

7. Wallis (1980), p. 322.

8. Wallis (1980), p. 323.

9. Friedman (1998), p. 145.

10. Wallis (1980), p. 329.

11. Tillman (2014), p. 13.

12. Tillman (2014), p. 13.

13. Tillman (2014), p. 250.

14. Wilson (2007), p. 3.

15. Hansen (2008), p. 286.

16. Hansen (2008), p. 286.

17. Hansen (2008), p. 279.

18. Wilson (2007), p. 1.

19. Hatfield (2003), p. 91.

20. Mangel (1984), p. 266.

21. Wald (1943).

22. Wald (1945).

23. Wald (1945), 1.1.

24. Wald (1945), 2.5.

25. Wald (1945), 4.2.

26. Wald (1945), B-1.42.

27. Ball (2003), p. 93.

28. Ball (2003), p. 99.

29. Wolfowitz (1952).

30. Shadrake (2014).

31. Shadrake (2014).

32. Shadrake (2014).

33. Illustrated War News (1915); Quinn (2020).

34. This example is from Pearl (2018), p. 343–46.

Chapter 6

1. The discussion of the origins of "History is written by the victors" is from Phelan (2019).

2. Phelan (2019).

3. Phelan (2019).

4. Gold (1997), p. 14.

5. Harris (2002), p. 40.

6. Harris (2002), p. 41.

7. Harris (2002), p. 44.

8. Harris (2002).

9. Harris (2002), p. 82.

10. Gold (1997), p. 41.

11. Gold (1997), p. 36.

12. Gold (1997), p. 81.

13. Gold (1997), p. 82.

14. Gold (1997), p. 92.

15. Pua (2019), p. 36.

16. Pua (2019), p. 35.

17. Pua (2019), p. 39.

18. Pua (2019), p. 37.

19. Drexler (2002), p. 245.

20. Pua (2019), p. 48.

21. Pua (2019), p. 46.

22. Pua (2019), p. 47.

23. Harris (2002), p. 87.

24. Description of "Cherry Blossoms at Night" from Geoghegan (2014) and Gold (1997).

25. Harris (2002), p. 52.

26. Harris (2002).

27. Harris (2002).

28. Harris (2002), p. 245.

29. Harris (2002), p. 246.

30. Harris (2002), p. 247.

31. Felton (2012), p. 92.

32. Harris (2002), p. 249.

33. Kinzer (2019), p. 27.

34. Harris (2002), p. 300.

35. Harris (2002), p. 302.

36. Kinzer (2019), p. 28.

37. Felton (2012), p. 102.

38. Harris (2002), p. 161.

39. Harris (2002), p. 160.

40. Gold (1997), p. 242.

41. Drayton (2005).

42. Felton (2012), p. 32.

43. Harris (2002), p. 397.

44. Kinzer (2019), p. 28.

45. Khan (2009).

46. Khan (2009).

47. Pawlowicz and Grunden (2015), p. 272.

48. Pawlowicz and Grunden (2015), p. 273.

49. Background on the decision to drop two atomic bombs on Japan comes from Huczko (2014), Ellsberg (2017), Frank (1999), and Alperovitz (1995). The role of the emperor and the Japanese government at the time comes from Bix (2000) and Laquerre (2013).

50. Huczko (2014), p. 134 and Ellsberg (2017), p. 261.

51. Ellsberg (2017), p. 262.

52. Frank (1999), p. 354.

53. Frank (1999), p. 345.

54. Frank (1999), p. 356.

55. Frank (1999), p. 347.

56. Huczko (2014), p. 133.

57. Huczko (2014), p. 131.

58. Huczko (2014), p. 131.

59. Frank (1999), p. 355.

60. Huczko (2014), p. 131.

61. Frank (1999), p. 303.

62. Frank (1999), p. 358.

63. Frank (1999), p. 303.

64. National Park Service (2017).

65. Hedges (2003).

66. Rummel (1997), p. 4. The term "megamurderers" was coined by Rummel.

67. Rummel (1997), p. 4.

68. Ding and Javed (2019).

69. Ding and Javed (2019).

70. Scheidel (2017), p. 225.

71. Rummel (1997), p. 100.

72. Rummel (1997), p. 97.

73. Rummel (1997), p. 105.

74. Li and Zhang (2013).

75. Yiwei (2013).

76. Johnson (2016).

77. Sherbakova (2019).

78. Scheidel (2017), p. 219.

79. Scheidel (2017), p. 220.

80. Rummel (1997), p. 81.

81. Rummel (1997), p. 81.

82. *Moscow Times* (2019).

83. Moscow Times (2019).

84. Moscow Times (2019).

85. Hemphill (1963).

86. Hemphill (1963), p. 224.

87. Hemphill (1963), p. 231.

88. Hemphill (1963), p. 401.

89. Williams (2012) is an example.

90. PBS.org (2021).

91. Diamond (1999), p. 348.

92. Diamond (1999), p. 203. A number of other studies have confirmed similar percentage declines. While the absolute number of Native Americans in the Americas before the arrival of the Europeans is still argued, there is agreement that the population was decimated by European diseases. See Ramenofksy (1987) and Alchon (2003) for a further analysis.

93. McNeill (1998), p. 216.

94. McNeill (1998).

95. Alchon (2003), p. 66.

96. McNeill (1998), p. 215.

97. Alchon (2003), p. 114.

98. Scheidel (2017).

99. Fenn (2001) describes the effects of various diseases, particularly smallpox, on the Indigenous inhabitants of the Americas.

100. McNeill (1998), p. 217.

101. McNeill (1998).

102. McNeill (1998), p. 26.

103. McNeill (1998).

104. Koch (2019).

105. S.J.Res. 14 (2009–2010).

106. S.J.Res. 14 (2009–2010).

107. Diaz (2017).

108. Diaz (2017).

Chapter 7

1. Background on Whiston's life and writings is from Force (2002) and Conover (2015).

2. Providentia (2019).

3. Swift (1898).

4. Swift (1898).

5. Descriptions of the effects of the comet or asteroid are a combination of Brusatte (2018), p. 314–315 and Brannen (2017), p. 173–218. Some scientists believe other factors also contributed to the Cretaceous mass extinction including a highly acidic ocean and the eruption of a chain of supervolcanoes that may have preceded the comet.

6. Steel (1995), p. 47.

7. Walsh (2019), p. 17.

8. Walsh (2019), p. 17.

9. Walsh (2019), p. 17.

10. The descriptions of dinosaurs come from Brusatte (2018). Where specific features are noted, a reference is provided.

11. Brusatte (2018), p. 336.

12. Brusatte (2018), p. 338.

13. Brusatte (2018), p. 198.

14. Brusatte (2018), p. 200.

15. Brusatte (2018), p. 204.

16. Brusatte (2018), p. 220.

17. Brusatte (2018), p. 219.

18. Everett (2017), p. 126.

19. Everett (2017), p. 126.

20. Everett (2017), p. 126.

21. Christian (2011), p. 165.

22. Descriptions of the Big Five mass extinctions are taken from Brannen (2017). There are many estimates for these numbers, and there is a degree of false precision in all of them concerning events that occurred so long ago. But scientists roughly agree on the order of magnitude.

23.   Finlayson (2010), p. 99.

24.   Hawks (2000), p. 9.

25.   Finlayson (2010), p. 99.

26.   Finlayson (2010), p. 99.

27.   Yost (2018).

28.   Pinker (2019), p. 294.

29.   Sterns (2000), p. x.

30.   Pinker (2019), p. 294.

31.   Bertram (2016) provides a model to quantify this qualitative statement.

32.   Scanes (2018), p. 84.

33.   Papagianni (2015), p. 21.

34.   Lee (2015), p. 176.

35.   Lee (2015), p. 184.

36.   Papagianni (2015), p. 13.

37.   Reich (2018), p. 26.

38.   Diamond (1999), p. 40 and Reich (2018), p. 28.

39.   Harari (2015), p. 145.

40.   Christian (2011), p. 175.

41.   Keeley (1996), p. 37.

42.   LeBlanc (2003), p. 97.

43.   Reich (2018), p. 40.

44.   Stringer (2012), p. 193.

45.   Shipman (2015).

46.   Stringer (2012), p. 204.

47.   Money (2019), p. 70.

48.   Brannen (2017), p. 226–233 for the species humans have driven to extinction.

49. Diamond (1999), p. 204.

50. Brannen (2017), p. 240.

51. Money (2019), p. 97.

52. Brannen (2017), p. 238.

53. Brannen (2017), p. 238.

54. Walsh (2019), p. 149.

55. Walsh (2019), p. 149.

56. Christian (2011), p. 142.

57. Like a cat perched on the windowsill of a high-rise apartment building, we believe our survival is not at risk. "After all," the cat may say to itself, "I have rested outside this open window for many years, peering down at the street fifty stories below, and nothing bad has happened." Unfortunately, cats who fell to their death are not around to reconsider whether a narrow ledge 600 feet above ground level is the best place to nap.

Chapter 8

1. Background on Neumann is from Macrae (1992), the definitive biography, and Poundstone (1992).

2. Macrae (1992), p. 24.

3. Gazzaniga (2018), p. 182.

4. Macrae (1992), p. 67–68.

5. Szanton (2003), p. 29.

6. Szanton (2003), p. 29.

7. Neumann (1955).

8. Neumann (1955).

9. Scharre (2018), p. 36.

10. Scharre (2018), p. 35.

11. Smil (2018), p. 369.

12. Smil (2018), p. 369.

13. Smil (2018), p. 372.

14. Scharre (2018), p. 38.

15. Keeley (1996), p. 54.

16. Christian (2011), p. 458.

17. Christian (2011), p. 458.

18. Christian (2011), p. 458.

19. Smil (2018), p. 365.

20. Christian (2011), p. 458.

21. Fazal (2019), p. 80.

22. Fazal (2019), p. 76.

23. Fazal (2019), p. 75.

24. Reagan (1983).

25. Graff (2018), p. 291–292.

26. Background on Operation Able Archer is from Jones (2016).

27. Background on nuclear false alarms is from Brimelow (2018) and Union of Concerned Scientists (2015). I did not include all the "close calls" or false alarms contained in the above. Rather, I have excluded those which in my view did not threaten global nuclear war such as accidents or the mishandling of nuclear materials.

28. Scharre (2018), p. 174.

29. Brimelow (2018).

30. Ellsberg (2017), p. 186.

31. Dobbs (2009), p. 6.

32. Dobbs (2009), p. 7.

33. Dobbs (2009), p. 21.

34. Joseph Kennedy made a fortune speculating in the stock markets during the roaring 1920s, before insider trading was illegal, and compounded that fortune by famously shorting stocks during the stock market crash of 1929. He then shrewdly invested these monies in stocks, real estate, and movie studios during the Great Depression, becoming one of the richest men in America by the 1940s. He also was the exclusive distributor of Gordon's Dry Gin and Dewar's Scotch in the United States after prohibition, but there is no evidence he was a "bootlegger," as has often been alleged.

    After WWII, he had hoped for a career in politics, first as governor of his home state of Massachusetts and perhaps afterward as president. But as ambassador to the United Kingdom from 1938 to 1940, he met numerous times privately with Hitler, was publicly anti-Semitic, and questioned the United States' support of Great Britain before Pearl Harbor. As a consequence, it was widely perceived he had been sympathetic with the Nazi regime, and his political ambitions were permanently dashed. He then devoted all his fortune to promoting his sons' political careers, including famously arranging some votes in Chicago to put Illinois in the column for his son Jack during the presidential election of 1960. JFK's margin of victory in the state was 8,858 votes, or 0.2 percent of the total votes cast. (It is worth noting JFK would have won the election, although just barely, without Illinois, but you can't blame a dad for making sure.)

35. Dobbs (2009), p. 22.

36. Dobbs (2009), p. 41.

37. Dobbs (2009), p. 41.

38. Dobbs (2009), p. 34.

39. Background on the events surrounding Soviet submarine B-59 comes from Dobbs (2009) and Ellsberg (2017).

40. Dobbs (2009), p. 302.

41. Ellsberg (2017), p. 217.

42. Ellsberg (2017), p. 217.

43. Ellsberg (2017), p. 217.

44. Ellsberg (2017), p. 212.

45. Lloyd (2002).

46. Ramseur (2017) describes the background on Maultsby's U-2 flight.

47. Dobbs (2009), p. 345.

48. Dobbs (2009), p. 353.

49. Arms Control Association (2003).

50. Perry (2015), p. 335.

51. Panda (2018).

52. Rosenbaum (2011), p. 234.

53. Duffin (2021).

54. Bracken (2013), p. 45.

55. Bracken (2013), p. 46.

56. Reagan (2011), kindle location 3697.

57. Carlin (2019), p. 150.

Chapter 9

1. Background on Fourier and his work is from Grattan-Guinness (2003) and *New World Encyclopedia* contributors "Joseph Fourier" (2021).

2. For a more detailed briefing on the greenhouse effect and its impact on global warming, see Houghton (2002).

3. Garner (2017).

4. Morton (2016), p. 67.

5. Brand (2010), p. 6.

6. Brand (2010), p. 6.

7. This is what most believe happened to Venus. As the largest CO_2 sink on the planet, the ocean boiled away and could no longer absorb CO_2, and eventually the air on Venus became what is it today, over 95 percent CO_2.

8. Morton (2016), p. 2.

9.   Billings (2013). The real answer is nobody knows. But the higher global temperatures get, the greater the risk that humans trigger a runaway greenhouse effect.

10.   Brand (2010), p. 60.

11.   Christian (2011), p. 346.

12.   Christian (2011), p. 406.

13.   Clark (2011).

14.   Diamond (2013), p. 475.

15.   Clark (2011).

16.   Clark (2011).

17.   Morton (2016), p. 3.

18.   Morton (2016), p. 3.

19.   Mann (2018), p. 27.

20.   Walsh (2019), p. 138.

21.   Walsh (2019), p. 138.

22.   Dwortzan (2016).

23.   Mann (2018), p. 73.

24.   Clark (2011).

25.   Clark (2011).

26.   Walsh (2019), p. 152.

27.   Grinspoon (2016), p. 204.

28.   Tollefson (2018).

29.   Clark (2011).

30.   Buck (2019), p. 123.

31.   Buck (2019), p. 124.

32.   Goodell (2010), p. 4.

33.   Morton (2016), p. 105.

34. Brand (2010), p. 284.

35. Goodell (2010), p. 17.

36. Mann (2018), p. 104.

37. Brannen (2017), p. 162.

38. Ward (2010), p. 210.

39. Grinspoon (2016), p. 188.

40. Falconer (2018), Lukacs (2012).

41. Chen (2018).

42. Chen (2018).

43. Welzer (2012), p. 36.

44. Dyer (2011), p. 58.

45. Dyer (2011), p. 206.

46. Dyer (2011), p. 206.

47. Welzer (2012), p. 46.

48. Ward (2010), p. 106.

49. Dyer (2011), p. 55.

50. Morton (2016), p. 121.

51. Dyer (2011), p. 58.

52. Dyer (2011), p. 59.

53. Ward (2010), p. 104.

54. Ward (2010), p. 106.

55. Dyer (2011), p. 77.

56. Gates (2021), p. 200.

57. Amadeo (2020).

58. National Institutes of Health (2015).

59. King (2020).

60. Let's face it, the most important technology behind the Apollo space program was television. I doubt the US would have gone to the moon, and certainly not more than once or twice, if the only publicity had been some newspaper articles.

Chapter 10

1. Background on Fermi is from Schwartz (2017) and Segre (2017).

2. Jones (1985). There are differing accounts of that day. The most definitive is Jones, as he corresponded directly with some of the principals and their written replies are included in his paper.

3. Bretscher (1955).

4. Levitin (2014).

5. There has been progress on one of the estimates required to resolve Fermi's paradox. Fermi and his fellow physicists at the time Fermi posed his question did not have a good guess at the number of Earth-like planets circling Earth-like suns. In 2009, NASA launched the Kepler satellite to survey the stars and planets of our galaxy. The results from that survey were published in 2013, and for the first time there is a good estimate of the number of Earth-like planets circling sun-like stars in the Milky Way: 11 billion (Grinspoon [2016], p. 340). This gives a starting point from which to speculate, a first known assumption, which Fermi and the other physicists did not have in 1950.

6. I have chosen to simplify the assumptions of Fermi's paradox compared with more complex formulations, such as Drake's equation.

7. Christian (2011), p. 95.

8. Christian (2011), p. 96.

9. Christian (2011), p. 97.

10. The example of the German tank problem is from The Albert Team (2020).

11. Ruggles (1947), p. 73.

12. Ruggles (1947), p. 73.

13. Davies (2006).

14. Davies (2006).

15. Ruggles (1947), p. 90.

16. Ruggles (1947), p. 88.

17. This is not strictly true, as the universe is expanding, but you get the general idea.

18. Walsh (2019), p. 296.

19. Fermi (2004), p. 142.

Chapter 11

1. Biographical details are from Gribbin (2013).

2. *New World Encyclopedia* contributors "Erwin Schrödinger" (2021), s.v. Middle years.

3. Gribbin (2013), p. 187.

4. Gribbin (2013), p. 193.

5. Bryne (2010), p. 144.

6. Ball (2018), p. 72.

7. Gribbin (2013), p. 180.

8. Ball (2018), p. 99.

9. Becker (2018), p. 56.

10. Becker (2018), p. 181 for a translation of the actual words Schrödinger used.

11. You are probably now a little bit concerned that this book has three stories about dead cats, but I assure you I have no problem with cats at all.

12. Gribbin (2013), p. 181.

13. Gribbin (2013), p. 182.

14. Gribbin (2013), p. 221.

15. Becker (2018), p. 14.

16. Ferris (1997), p. 276. Again, I have nothing against cats.

17. The arguments over Goldilocks universes are usually related to a discussion about the multiverse and big bang. However, I see no reason that the same speculations could not apply to the MWI.

18. Davies (2006), p. 146.

19. Davies (2006), p. 146.

20. Harris (1991), p. 51.

21. Barrow (2009), p. 266.

22. Thomas (2017), p. 83.

Index

T

taxation, lotteries as form of, 43
technological innovation, 173–75
technology, intrinsic danger of,
      125–28
Think and Grow Rich (Hill), 30–33
threat of nuclear war, reducing,
      148–50
Toba eruption, 115
Toft, Mary, 108–9
tourniquets, use on battlefield of,
      80–81
tragedy of the commons, 162
trauma care, 55–56
*True and Faithful Narrative of What
      Passed in London, During the
      General Consternation of All
      Ranks and Degrees of Mankind;
      on Tuesday, Wednesday,
      Thursday, and Friday Last, A*
      (Swift), 110
Truman, Harry S., 91–95
Tyrannosaurus rex, 112–13

U

Unit 731, 84–91
United States
      biological warfare program, 88–91
      bombing of Hiroshima and
            Nagasaki, 91–95
      Cuban Missile Crisis, 139–44
      false alarms of nuclear attacks,
            138–39
      genocide of Native Americans,
            99–103
      global warming, effect on, 172–73
      Operation Able Archer, 136–38
      Paris Agreement, 161–62
      Positive Security Assurance,
            146–47
      slavery in, 98–99
      threat of nuclear war, reducing,
            148–50
      tourniquet use on battlefields,
            80–81
      WWII aircraft vulnerabilities
            analysis, 70–79

V

V-1 rocket production, 188
Venus, 156
von Neumann probes, 193–94

W

Wald, Abraham, 67–70, 80–81
      aircraft vulnerability analysis,
            71–79
WAP (weak anthropic principle),
      208–10
warfare, 65. *See also* nuclear war
      Abraham Wald, 67–70
      bombing of Hiroshima and
            Nagasaki, 91–95
      estimated deaths from, 130
      German tank problem, 187–88
      growth in power of weapons,
            128–31
      Japanese biological warfare
            program, 84–91
      tourniquets, 80–81
      WWII aircraft vulnerabilities,
            70–79
      WWII helmets, 79–80
Waterman, Robert, 25–28
weak anthropic principle (WAP),
      208–10
weapons. *See also* nuclear war
      bombing of Hiroshima and
            Nagasaki, 91–95
      growth in power of, 128–31
      Japanese biological warfare
            program, 84–91
weight loss industry, 57–60
*Where Are the Customers' Yachts?
      or A Good Hard Look at Wall
      Street* (Shwed), 10–12, 22–23
Whiston, William, 107–10
Wigner, Eugene, 126
wild animal species, extinction of,
      120–21
Winthrop, John, 101–2
Wolfowitz, Jacob, 78–79
World War I, 129–31

About the Author

DAVID LOCKWOOD is a former lecturer on the faculty of the Graduate School of Business at Stanford University. He has three decades of experience working on Wall Street and in Silicon Valley and has served on more than twenty public and private company boards. In addition, he has been a senior advisor to the US Department of Energy. He currently resides in the Western Rockies with his wife and three children.